ESSENTIAL
INVERTEBRATE
ZOOLOGY

ESSENTIAL

INVERTEBRATE ZOOLOGY

M. S. Laverack

BSc, PhD, FRSE
Professor of Marine Biology and
Director, Gatty Marine Laboratory,
University of St Andrews, Scotland

J. Dando

BSc, MSc
Research Associate,
University of St Andrews, Scotland

A HALSTED PRESS BOOK

JOHN WILEY & SONS

NEW YORK

© 1974 Blackwell Scientific Publications
Osney Mead, Oxford,
3 Nottingham Street, London W1M 3RA,
9 Forrest Road, Edinburgh,
P.O. Box 9, North Balwyn, Victoria, Australia.

All rights reserved. No part of this publication
may be reproduced, stored in a retrieval system,
or transmitted, in any form or by any means,
electronic, mechanical, photocopying, recording
or otherwise without the prior permission of
the copyright owner.

Library of Congress Cataloging in Publication Data

Laverack, M S
    Essential invertebrate zoology.

    "A Halsted Press book."
    1. Invertebrates. I. Dando, J., joint author.
II. Title. [DNLM: 1. Invertebrates. QL362 L399e]
QL362.L29   592   74-8619
ISBN 0-470-51887-1

First published 1974

Published in the United States of America by
Halsted Press
Division of John Wiley & Sons, Inc.
New York

Printed in Great Britain

CONTENTS

	Apologia	vii
	Acknowledgments	viii
1	Generalities	1
2	Phylum PROTOZOA	6
3	Phylum MESOZOA	15
4	Phylum PORIFERA	18
5	Phylum CNIDARIA	22
6	Phylum CTENOPHORA	31
7	Phylum PLATYHELMINTHES	34
8	Phylum NEMERTINI	44
9	Phylum GASTROTRICHA	49
10	Phylum KINORHYNCHA	52
11	Phylum ROTIFERA	54
12	Phylum PRIAPULOIDEA	57
13	Phylum NEMATOMORPHA	60
14	Phylum ACANTHOCEPHALA	63
15	Phylum NEMATODA	66
16	Phylum ENTOPROCTA	71
17	Phylum ANNELIDA	74
18	Phylum ECHIURIDA	87
19	Phylum SIPUNCULIDA	93
20	Phylum POGONOPHORA	97
21	Phylum ONYCHOPHORA	102
22	Phylum ARTHROPODA	105
23	Phylum TARDIGRADA	132
24	Phylum PENTASTOMIDA	135
25	Phylum MOLLUSCA	138
26	Phylum PHORONIDA	153
27	Phylum ECTOPROCTA=(BRYOZOA)	157
28	Phylum BRACHIOPODA	162
29	Phylum CHAETOGNATHA	166
30	Phylum ECHINODERMATA	168
31	Phylum UROCHORDATA	179
32	Phylum HEMICHORDATA	185
33	Phylum CEPHALOCHORDATA	189
	Systematic Index	193
	Index of Genera	195

Apologia

It is admittedly difficult to do justice to the 30 or so phyla, and numerous classes, that comprise the invertebrates, in the short space available in this book. Nevertheless there is a case to be made for the compression of the detail given in the standard texts available on the invertebrate groups, though this compression should not be so great as to lead to incomprehension.

The impetus of modern biology has been, for the last fifty years or more, towards an experimental and analytical understanding of living creatures and their adaptability to innumerable modes of life. The development of cell biology, endocrinology, ecology, parasitology, comparative physiology, ethology, neurophysiology, and other scientific disciplines, has led to both synthesis and fragmentation of approach.

Fragmentation has occurred for the reason that the older comparative anatomy studies that provided the foundation of descriptive biology, have gradually lost their unifying pre-eminence, to be replaced by a series of other subjects within subjects. These newly-derived fields are still in the process of development, but show the way that biology will progress in the future.

Synthesis from these subjects arises because there is a growing enthusiasm for a viewpoint that encompasses all aspects of the economy of animals. Instead of a narrow approach that accepts only morphology as the yardstick and cornerstone there is a growing attempt to understand animals as metabolising, behaving, distributed populations. Anatomy and morphology are but two aspects of their being.

As a consequence of these evolutionary trends in zoology more and more university departments are catering for the divisive elements, with short courses in many subjects, and with the introductory stages of anatomical and morphological studies decreasing in length. Let no one doubt, however, that a sound grounding in the biology of an animal group depends upon an introduction to the body form and organ function of the animals themselves.

This book attempts such a task. If the time available for a consideration of the invertebrates is short, say 12—15 weeks, then the basic detail necessary is rather less than that available in most texts. Such books are still invaluable for those with time to study them, but for others the question is "What is the kernel of knowledge for each group?" We have attempted to condense the known material into a concise form. It will not satisfy the senior student of any group, and many important groups (e.g. the insectan orders) are omitted. Who, in 12 weeks, can learn them all and know the characters, and for what purpose, if it is only to forget them in the subsequent weeks of study of cell nuclei, action potentials, or hormones?

We fully realize that many criticisms may be levelled at the format, at the material included, and at that omitted. If, however, we have provided sufficient to act as a guide to students, and better still, to release the teacher from the drudgery of simply reproducing the anatomy, we hope to have created a platform for profitable expansion on innumerable other topics. If so, our job is well done.

Acknowledgments

We are indebted to many colleagues for their help, but especially to the following who read various parts of the manuscript, and corrected our errors:

Drs J. B. Tucker,
    A. C. Campbell,
    M. Crisp,
    I. D. McFarlane,
    M. R. Dando.

Any faults that remain are entirely those of the authors.

We also wish to thank Mr John Stevenson for technical assistance and Mrs Doris Hunter for typing the manuscript. Andrea Sylvester also helped in many ways.

We are also grateful to the following for permission to use extracts from their published works:
Dr E. C. Southward and George Allen and Unwin Ltd;
Professor R. B. Clark and Academic Press Ltd.

Finally we are grateful to the following publishers and authors for permission to use figures from their various publications as indicated below:
Academic Press Inc. (copyright holders), and A. Bidder for Fig. 4i, Chapter 3, *Physiology of Mollusca*, Vol. 2; and G. Jagersten for Figs. 21, 25C, 39, *Evolution of the Metazoan Life Cycle*; and A. W. Martin for Fig. 1, Chapter 11, *Physiology of Mollusca*, Vol. 2; and J. E. Morton for Figs. 17, 18, Chapter 12, *Physiology of Mollusca*, Vol. 1; and G. Owen for Figs. 4.2B, 5, Chapter 2, *Physiology of Mollusca*, Vol. 2; and M. Yonge for Figs. 2A, 9, 11, 12, 14, 20, Chapter 1, *Physiology of Mollusca*, Vol. 1; B. Mulloney for Fig. 1, *Science*, 168, 994–996, copyright 1970 by the American Association for the Advancement of Science; British Museum (Natural History) and A. M. Clark for Figs. 1, 2, 10, 30, *Starfishes and their allies*; Company of Biologists and R. A. Hammond, for Fig. 3, *J. Exp. Biol.* 45, 203; and K. Johansen & A. W. Martin for Fig. 3, *J. Exp. Biol.* 43, 333; and J. V. Lawry for Fig. 1, Paper 1; Fig. 12, Paper 2; *J. Exp. Biol.* 45, 1966; Hutchinson & Co. and R. Dales for Fig. 18, *Annelids*; and R. Gibson for Fig. 6, *Nemerteans*; and D. Nichols for Fig. 17, *Echinoderms*; and M. J. S. Rudwick for Figs. 3, 48, 79, *Brachiopods, Living and Fossil*; and J. S. Ryland for part of Fig. 7, 13 and Fig. 3, *Bryozoans*; Macmillan Publishing Co. Inc. and W. D. Russell-Hunter for Figs. 6.1, 6.2, Vol. 1; Figs. 2.3A, 4.1, 4.2C, 4.3B, 9.2, 13.3, Vol. 2, *A Biology of Lower Invertebrates*; Marine Biological Association of the United Kingdom and E. J. Denton for Fig. 7, *J. mar. biol. Ass.* 41, 365; The Ray Society and A. Graham for Fig. 246, *British Prosobranch Molluscs*; Springer-Verlag and P. R. Flood for Fig. 1, *Z. Zellforsch*, 103, 115; Weidenfeld and Nicholson and M. J. Wells for Figs. 6.9, 10.2B, *Lower Animals*, World University Library; John Wiley & Sons Inc. and D. Nichols for Figs. 9.1, 9.2, 9.4, Chapter 9, *Physiology of Echinodermata*: Interscience; The Zoological Society of London and R. A. Hammond for 3 Figs. from *J. Zool.* 162, 469.

1 Generalities

The invertebrates are not a homogeneous group of phyla. Rather they are a miscellany of animals, falling into a number of phyla, some of which are evidently related whilst others are not. In this they show a much greater range of form and adaptations than do the vertebrates.

The plan of this book is to deal with the phyla in an orderly fashion with closely related groups, so far as they can be ascertained, treated in succession. The very fragmentation of such an approach, however, denies the opportunity of dealing with topics that recur throughout the phyla to greater or lesser degree. In the next few pages we propose to outline some of the salient features that are observed amongst the invertebrates; some key words are defined at the end of this section.

1 Symmetry

Four kinds of symmetry are possible, and all are found amongst invertebrates. The concept of symmetry allows for the division of a whole into 2 or more equal portions, by separation along lines or planes.

a *Asymmetry*: there are no planes of regular symmetry, e.g. in protozoans such as rhizopods.

b *Spherical symmetry*: the body is divisible into symmetrical halves in all directions and planes; found in protozoans (Radiolaria).

c *Radial symmetry*: in which lines of symmetry exist, but in the vertical planes only; typical of phyla of sessile or floating animals, namely Cnidaria and Echinodermata.

d *Bilateral symmetry*: found in freely mobile animals with consequent development of dorsal and ventral surfaces, and of anterior and posterior ends; only one plane of symmetry divides the animal into symmetrical halves; includes the majority of active invertebrates.

2 Cephalization

An animal exhibiting radial symmetry presents a similar attitude to the environment around the circumference. It is identical no matter from which direction it is approached. Development of bilateral symmetry in active animals was concomitant with movements concentrated in a particular direction, thence the animal body polarized with one end leading. The concentration of sense organs, and usually the feeding apparatus, occurred at the leading end, and this is the process of cephalization which is associated with the specialization of a portion of the nervous system known as the brain, and the formation of a head.

3 Coelom

The increasing complexity of the animal body as it is followed from the unitary Protozoa through the simple Mesozoa to the lower and then more advanced Metazoa, has led to increasingly sophisticated physical apparatus, evolved under adaptive pressures for greater efficiency and metabolic economy.

The increase in size and development of musculature of more advanced groups led to greater independence from the environment and more powerful locomotion became possible. Cilia and poor muscle performance are sufficient only for floating, sessile or sluggish animals. Development of a strong, incompressible, but flexible and not solid skeleton enabled great improvements to occur in mechanical activity.

Invertebrates (ignoring the Protozoa, Porifera and diploblastic Cnidaria and Ctenophora) may be classified as (Table 1.1)

Table 1.1 Metazoa.

Acoelomate	Pseudocoelomate	Coelomate
Platyhelminthes	Acanthocephala	Annelida
Nemertini	Rotifera	Mollusca
	Gastrotricha	Arthropoda
	Kinorhyncha	Priapuloidea
	Nematoda	Ectoprocta
	Nematomorpha	Phoronida
	Entoprocta	Brachiopoda
		Chaetognatha
		Echinodermata
		Hemichordata
		Urochordata
		Cephalochordata

(i) acoelomate: the body volume is filled with mesenchyme and other tissue
(ii) pseudocoelomate: with a body space which is fluid-filled but not formed from mesoderm and not lined by a cellular peritoneum
(iii) coelomate: with a fluid-filled body cavity formed from mesodermal sources, originating either from mesoderm blocks or sacs of the embryological enteron wall. This is much modified in some groups (e.g. Mollusca, Arthropoda) and details are given in appropriate chapters.

Embryologically the coelom derives either from
I the cavitation of blocks of mesoderm: schizocoely; or
II the formation of sacs of mesoderm from the archenteron: enterocoely.

4 *Metamerism*

The phenomenon of metamerism is equated with the meristic segmentation of a bilaterally symmetrical elongate body. Each segment contains one pair of some or all of the organs of the body (nephridia, coelomoducts, gonads, ganglia). Metamerism appears in the annelids, arthropods, the chordates (not dealt with in this book) and the cestodes. Most cestodes form many identical segments that are shed terminally when the reproductive capacity is fulfilled. Annelids appear to have developed septate segmentation and compartmentation of the nervous and muscular systems under the adaptive stress of locomotory and burrowing needs.

5 *Cleavage*

The embryology and development of invertebrates takes place from a variety of egg types, through a number of styles of division to form a number of larval types and juveniles. There are, however, certain strong lines of similarity between groups.

a *Egg types*: cleavage patterns depend partly upon the nature, size and relative proportions of the egg, as shown in Table 1.2.

Table 1.2 Types of egg and cleavage.

Egg type	Amount of yolk	Cleavage	Type
i. Isolecithal	small or none	holoblastic	total cleavage
ii. Telolecithal	large	meroblastic	only animal pole divides

b *Cleavage*: the division of the fertilized egg follows two major patterns amongst metazoan invertebrates. Both are holoblastic and involve the whole egg; they are termed radial and spiral cleavage (Fig. 1.1; Table 1.3).

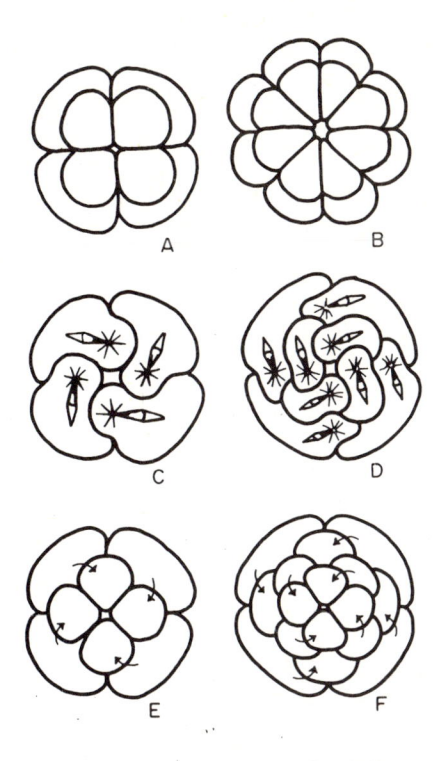

Fig. 1.1 Comparison of radial and spiral cleavage. A and B. 8- and 16-cell stages of radial type. C and D. 3rd and 4th cleavages of spiral type. E and F. 8- and 16-cell stages, spiral type (A—F after Richards).

Table 1.3 Patterns of egg division.

Radial	Spiral
Plane of cleavage is always at right-angles or parallel to the polar axis	Planes of cleavage lie at an angle to the polar axis, so that any one cell lies betwixt two others above or below it

Another radical difference exists between these two cleavage styles. In spirally-cleaving eggs, by the time 32 cells have been formed, the future position and function of each blastomere is fixed. This is determinate development. Radially-cleaving eggs do not show such fixation of cellular fate, each cell retaining the ability to form various kinds of organs and tissue until late in development. This is the indeterminate type.

6 *Deuterostomia — Protostomia*

The two lines of development of invertebrate groups outlined above are paralleled by other features that characterize these lines. This is taken as evidence for evolutionary trends among the phyla and for relationships between groups. It must, however, be stressed that many small phyla have intermediate characters falling between the two major projected lines, and that not all representatives lying within these two lines in fact exhibit all the characteristic features. There is thus room for discussion as to the validity of assuming evolutionary trends based on these factors.

Nonetheless it is worthwhile indicating the similarities that are claimed (Table 1.4).

Table 1.4 Protostomia—Deuterostomia groupings.

Line	Protostomia	Deuterostomia
Phyla	Annelida Arthropoda Mollusca Platyhelminthes Nemertini	Echinodermata Chordata Hemichordata
Characters 1	spiral cleavage	radial cleavage
2	blastopore forms mouth	blastopore forms anus (mouth is new structure)
3	coelom schizocoelic	coelom enterocoelic and often tripartite
4	CNS ventral	CNS dorsal or superficial
5	trochophore larva	dipleurula larva

Generalities

KEY WORDS

Acoelomate   lacking a coelom.

Appendage   a large (relative to animal size) projection from the body, e.g. parapodia in annelids, limbs of arthropods.

Blastopore   when an embryo gastrulates, the primitive internal cavity (archenteron) communicates with the exterior via the blastopore. The fate of the blastopore is significant in classification of **Deuterostomia** (where the area forms the anus) and **Protostomia** (in which it forms the mouth).

Blastula   form of an embryo just before gastrulation; usually a ball of cells, surrounding a fluid cavity.

Botryoidal tissue   the packing tissue of the leech body.

Brain   anterior portion of the centralized nervous system, well developed in cephalization and aggregation of sense organs anteriorly in bilaterally symmetrical animals.

Central nervous system (*CNS*)   name given to the concentrated regions of nervous tissue that contain ganglia (nerve cell collections) which are linked together by nerve trunks (containing interneurones) and from which originate the peripheral nerves (carrying motor fibres to effectors and sensory fibres from sense organs).

Cephalization   process leading to the formation of a distinctive head region.

Chaetae (*setae*)   bristles of chitin, characteristic of annelids, that project from the animal surface.

Cilium   short motile structure at cell surface. Has an internal structure remarkably consistent throughout the animal kingdom. In some cases, such as sense organs, this structure is modified.

Cirrus   a slender, flexible appendage.

Class   a grouping of similar orders.

Coelom   the main body cavity of triploblastic animals. Lined by epithelium and situated in the mesoderm from which it is usually formed by either schizocoelic or enterocoelic processes. Fluid-filled, often with migratory cells, and gametes formed within it. Various other terms are derived from the same root meaning cavity, e.g. spongocoel (sponges), axocoel (echinoderms).

Coelomoduct   a passage linking the internal coelom with the external environment. Frequently used for passage of excretory and/or gonadal products. Lined by mesodermal tissue.

Colonial   group of animals which are associated and usually (but not invariably) physically linked together.

Cuticle   epidermally secreted non-cellular layer of chitin or collagenous protein.

Diploblastic   body composed of ectoderm and endoderm, e.g. coelenterates.

Dorsal   that side of the body normally directed upward with regard to gravity. May, however, differ with stage of development as in echinoderms, where metamorphosis is complex and resultant adult postures do not represent normal dorsal, ventral, right and left sides.

Enterocoely   process by which coelom derives embryologically from pouches of the archenteron.

Family   a classificatory group, composed of a number of similar genera.

Flagellum   long filamentous projection from a cell surface. Characteristically undergoes undulatory or sinuous movements.

Flame cells   also known as solenocytes. Cells with a lumen containing cilia that beat as a group. Many may be connected together by canals opening eventually to the exterior. Believed to be involved in water regulation activities.

Ganglion   a collection of nerve cell bodies and associated neuropile.

Gastrulation   the process by which the blastula develops into the gastrula. The wall of the blastula invaginates forming the internal archenteron bounded by the endoderm, the exterior layer being the ectoderm, and with the subsequent formation of the third primary tissue layer the intermediate mesoderm.

Gill   respiratory surface of aquatic animals.

Gill slit   opening joining the exterior with the interior of the pharynx (see Hemichordata).

Gland   an organ or cell that synthesizes specific materials (secretions) that may be liberated internally (endocrine) or externally (exocrine).

Haemocoel   a large blood space. Continuous with and part of, the blood system, but fulfilling the structural function of a coelom in some groups where it is much increased in volume. It never communicates directly with the outer environment, and germ cells are not found in it.

Holoblastic   in embryology indicates that the whole egg undergoes cleavage (usually associated with small yolk content).

Larva   the immature pre-adult form. Occurs in many invertebrate groups. Usually feeds, but in a different way to the adult.

Mantle   also known as the pallium. A fold or flap of

tissue that covers the whole or part of the body. Secretes shell where present, or covers vulnerable organs by being thickened and coloured. Found in molluscs, and a similar organ is present in brachiopods.

Meroblastic in embryology describes condition in which only part of the egg undergoes cleavage, as in large, yolky eggs.

Metamerism repetition of segments, each containing a more or less full complement of organs. May be considerably modified.

Metamorphosis the process, often dramatic, by which the larva attains adult form.

Morphology the study of shape and form.

Nauplius early larval type of many Crustacea; bears three pairs of appendages.

Nephridium a tube of ectodermal origin, opening to the exterior at one end, and which may be open to the coelom or closed (by flame cells) internally. When the tube is combined with a coelomoduct the resultant structure is a nephromixium. Probably involved in excretory and regulatory processes.

Notochord overall term for what now appears to be a structure of doubtful homology in Hemichordata, Cephalochordata and Urochordata amongst invertebrates. Lies dorsally between gut and nervous system. Composed of muscle in Cephalochordata, and vacuolated cells in the other two groups.

Ocellus simple light receptor.

Ommatidium one of the component parts of a compound eye typical of arthropods.

Order one of the commonly used groups in classification. Consists of a number of related families.

Ostium an aperture, as in the heart of Crustacea, or the walls of sponges.

Ovoviviparous embryos developing within the maternal parent, and though the egg membranes may persist, they obtain nutrient from the parent.

Parenchyma tissue that provides a loosely connected packing of the body space in some groups.

Phylum consists of one or a number of classes in the taxonomical arrangement of animals.

Pseudocoel in some groups of invertebrates the body space is filled with vacuolated cells of gelatinous consistency, others have small fluid filled spaces not thought to be lined by mesodermally derived epithelia. These are pseudocoels.

Receptor specialised cells or collections of cells that are concerned with monitoring changing conditions. May be internal or external.

Schizocoely process by which coelom originates as cavities formed by splitting of mesodermal blocks.

Sessile fixed to the substratum.

Seta bristle epidermally produced, composed of solid cuticle alone (annelids) or cuticle with a lumen (insects).

Statocyst a vesicular organ containing statoliths (sand grains or secreted calcified platelets) acting as a gravity detector.

Triploblastic body made up of three primary layers: ectoderm, mesoderm and endoderm.

Trochophore (*trochosphere*) ciliated, planktonic larva with a main band of cilia equatorially around the body anterior to the mouth.

Veliger modified trochophore of the molluscan group.

Ventral that side of the body which is normally directed downward with respect to gravity.

Zooid member of a colony of animals which are joined together.

2 Phylum PROTOZOA

About 30,000 species are known of which the majority are microscopic.

Characteristics

1 With the exception of cysts, all are confined to moist places.
2 Single-celled or acellular organisms which combine normal cellular functions with those like food capture, locomotion etc. for which higher organisms possess special tissues or organs.
3 Lack tissues and organs.
4 The functionally differentiated parts are the organelles.
5 A period of encystment is commonly part of the life-cycle.

Larval form

Larval forms, as present in other phyla, do not occur. However, amongst a few ciliates in which the adult is ectoparasitic or sessile and lacks somatic ciliature, there may be a ciliated free-swimming distributive phase. This 'larva' is produced from the entire adult body or by fission or budding, i.e. it does not result from sexual reproduction.

Adult body form

The cell possesses one or more nuclei bounded by nuclear envelopes, a cell membrane bounding the cytoplasm (which may be differentiated into outer ectoplasm and inner endoplasm, or outer cortex and endoplasm), and generally, an endoplasmic reticulum and mitochondria. It may also possess Golgi apparatus, centrioles, chromoplasts, cilia, flagella etc. The cytoplasmic membrane, the plasmalemma, may be augmented by other layers to form a stouter living pellicle in some species. A cuticle, shell, exo- or endoskeleton may be present.

1 *Class* MASTIGOPHOREA (Flagellata) e.g. *Cryptomonas, Euglena, Ceratium, Monas, Trypanosoma, Trichonympha.* With one or more flagella during most of the life-cycle; usually with a pellicle or cuticle; lacking a macronucleus; includes both plant-like (phytoflagellate) and animal-like (zooflagellate) forms (Fig. 2.1); with some colonial species (e.g. *Volvox*); parasitic forms are rarely intracellular.

2 *Class* OPALINATEA e.g. *Opalina*, (Fig. 2.3A) *Zelleriella.* Densely ciliated, possessing more than one nucleus, all of similar type; with a pellicle. Differ from ciliates in single type of nucleus, form of the infraciliature, sexual syngamy, and occurrence of division plane running between ciliary rows. Differ from flagellates in having more than 1 nucleus and lack of centrioles in division.

3 *Class* SARCODINEA (Rhizopoda) e.g. *Amoeba, Entamoeba, Difflugia, Acanthometra, Actinosphaerium.* Possessing one or more pseudopodia and lacking flagella during the principal, amoeboid phase of the life-cycle (Fig. 2.2); naked or possessing an external shell, or with a central capsule and endoskeleton of spicules; lacking a macronucleus; with a few parasitic forms.

4 *Class* CILIOPHORA (Ciliata) e.g. *Paramecium, Stentor, Vorticella, Dendrocometes.* Containing the most highly-organized protozoans; never amoeboid but possessing cilia, at least during young stages; typically with a pellicle, sometimes with a chitinous case or cuticular cup; generally with a mouth or cytostome; typically possessing two types of nuclei and usually with a macronucleus; free-living and parasitic forms (the latter are never intracellular); most species solitary and motile but there are some sessile and colonial forms (Fig. 2.3).

5 *Class* SPOROZOA e.g. *Eimeria, Plasmodium, Monocystis, Helicosporidium, Haplosporidium.* In the principal phase of the life-cycle lack locomotory

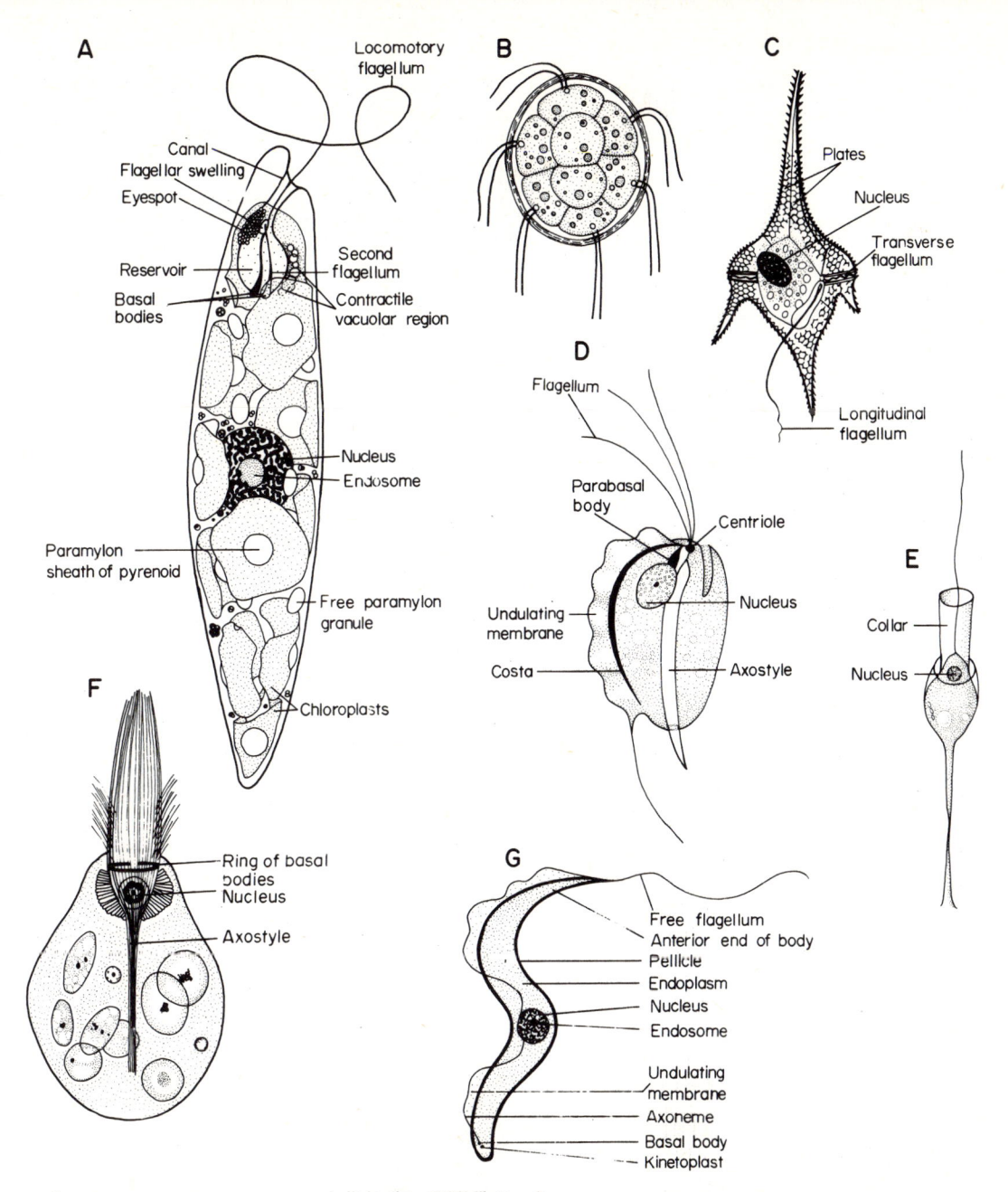

Fig. 2.1 Mastigophorea (A—C phytoflagellates, D—G zooflagellates). A. *Euglena gracilis* (after Leedale). B. *Pandorina*, a colonial phytoflagellate (after Hyman). C. *Ceratium* (after Stanier, Doudoroff and Adelberg). D. *Trichomonas* (after Grell). E. *Codosiga* (after Lapage). F. *Lophomonas* (after Kudo). G. *Trypanosoma* (after Bullough).

Phylum PROTOZOA

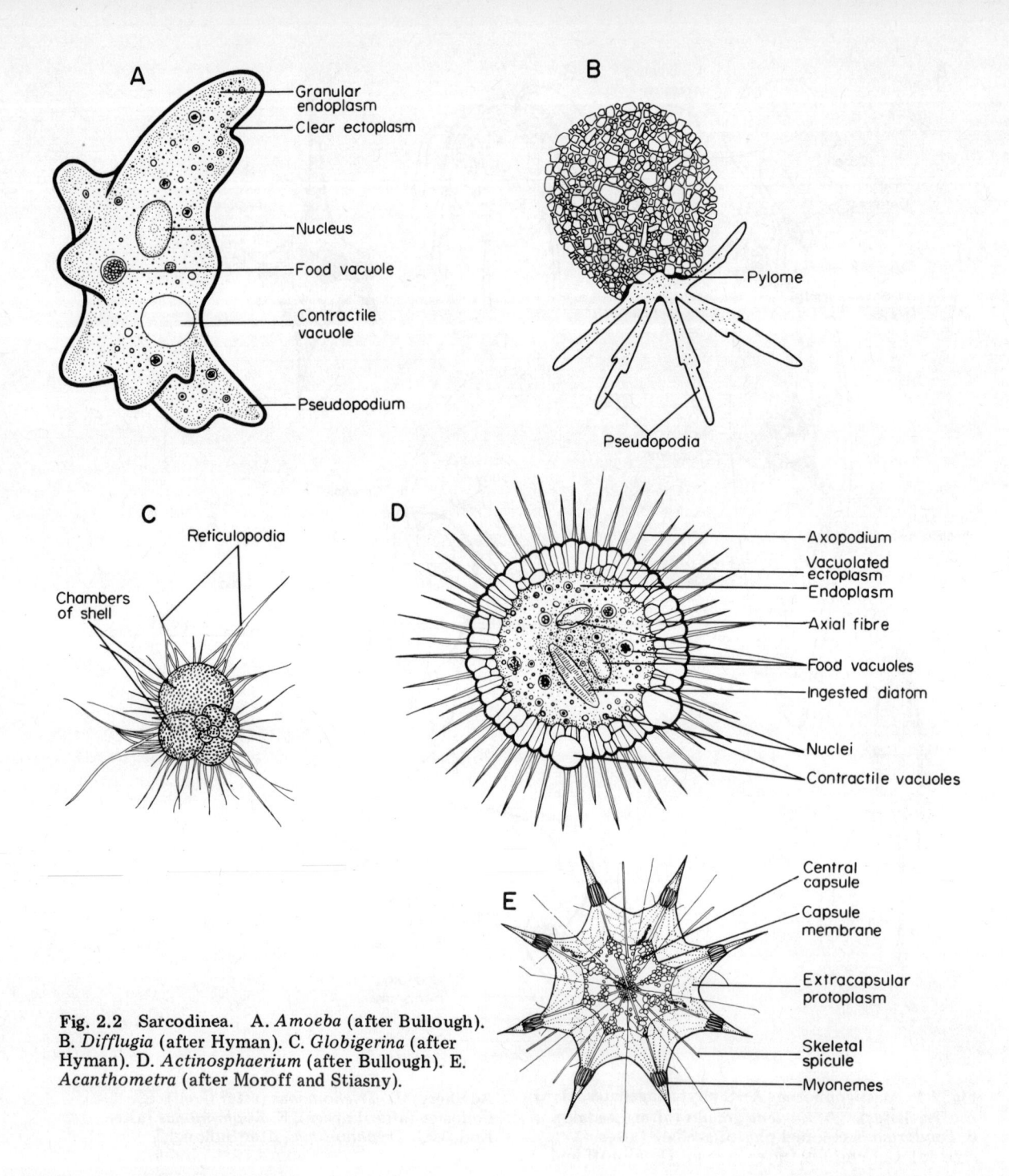

Fig. 2.2 Sarcodinea. A. *Amoeba* (after Bullough). B. *Difflugia* (after Hyman). C. *Globigerina* (after Hyman). D. *Actinosphaerium* (after Bullough). E. *Acanthometra* (after Moroff and Stiasny).

Phylum PROTOZOA

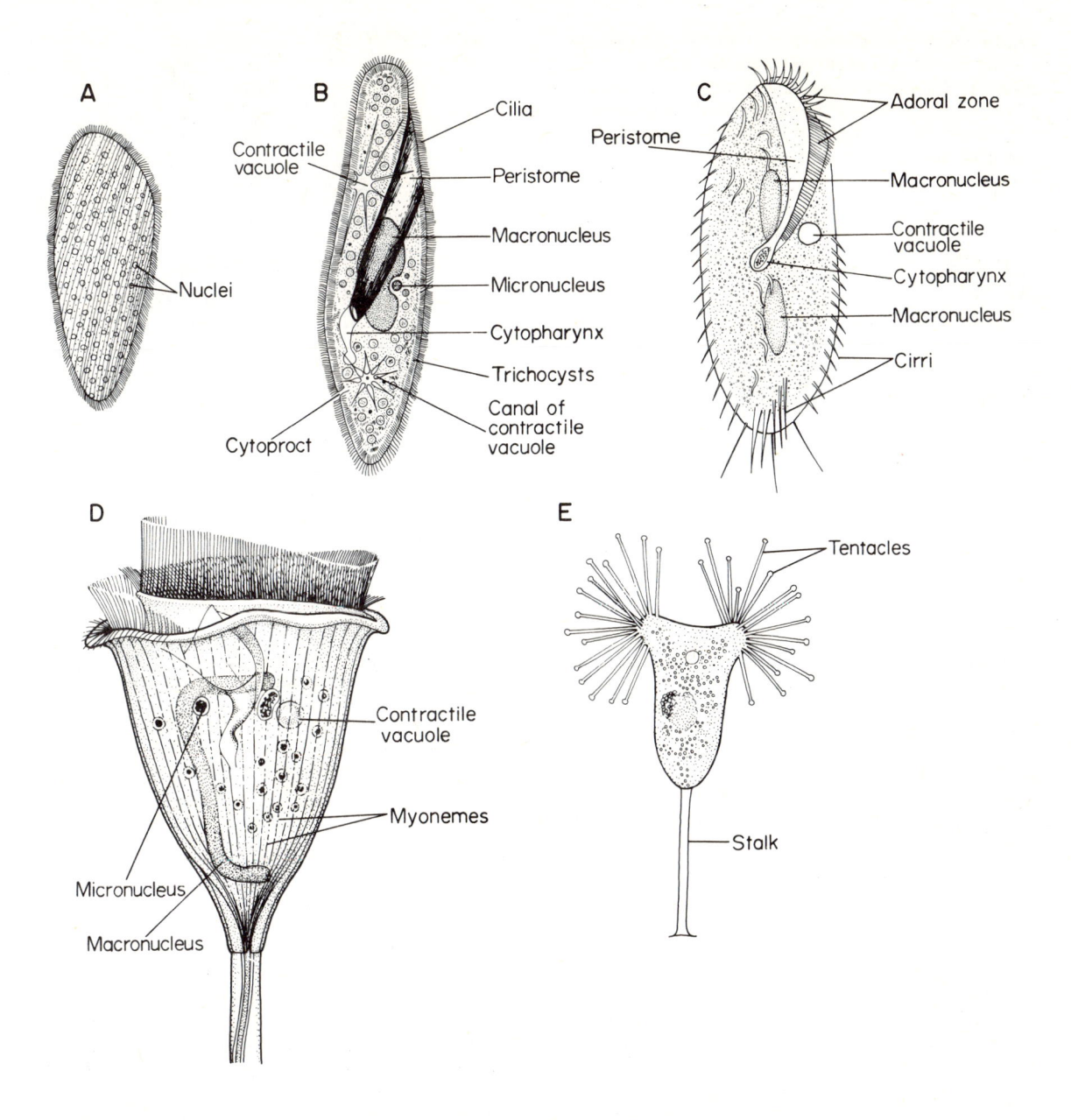

Fig. 2.3 Opalinatea. A. *Opalina*, a multinucleate species (after Hyman) Ciliophora. B. *Paramecium caudatum* (after Hyman). C. *Stylonychia* (after Hyman). D. *Vorticella*, a stalked species (after Grell). E. *Acineta*, tentaculate adult (after Kozloff).

organelles or are amoeboid; naked or with a pellicle having micropores at the cell surface; lacking a macronucleus; producing large numbers of spores after syngamy; internal parasites of other animals (Fig. 2.4); usually with an intracellular stage; with complex life-cycles (Figs. 2.5, 2.6).

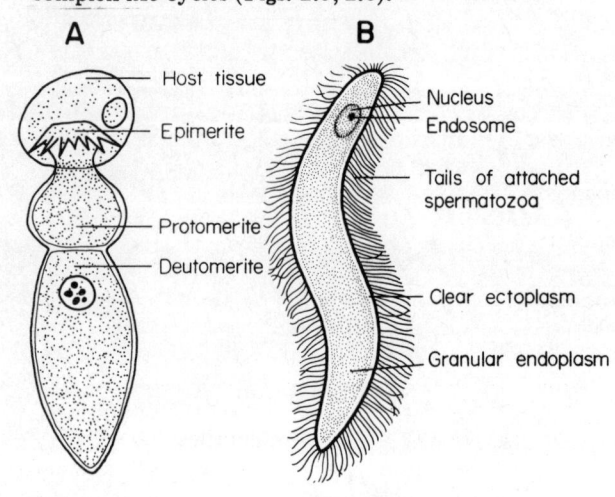

Fig. 2.4 Sporozoa. A. *Actinophilus*, a cephaline gregarine (after Hyman). B. *Monocystis*, an acephaline gregarine often found in the seminal vesicles of earthworms (after Bullough).

6 *Class* CNIDOSPORA e.g. *Myxobolus, Nosema, Ichthyosporidium*. Small amoeboid organisms; producing spores either multicellular with one or more sporoplasms and filaments, or unicellular and unifilamentous; undergo multiple spore formation in a host which they parasitize.

Feeding

a The phylum contains both autotrophic and heterotrophic species. Autotrophs are able to synthesize complex organic compounds from simple inorganic substances using energy obtained from sunlight (phototrophs, e.g. phytoflagellates with chromoplasts) or from chemical reactions (e.g. some colourless phytoflagellates). Heterotrophs utilize only materials manufactured by other organisms, either absorbing them through the body surface (osmotrophs, e.g. some parasitic species) or ingesting solid particles (phagotrophs, e.g. many sarcodines and ciliates). Phagotrophs may be filter-feeders or may engulf large particles, some are raptorial.

It seems that many protozoans are not restricted to one mode of feeding, e.g. some autotrophic phytoflagellates may require simple organic compounds; some protozoans may be osmotrophs or phagotrophs at different times.

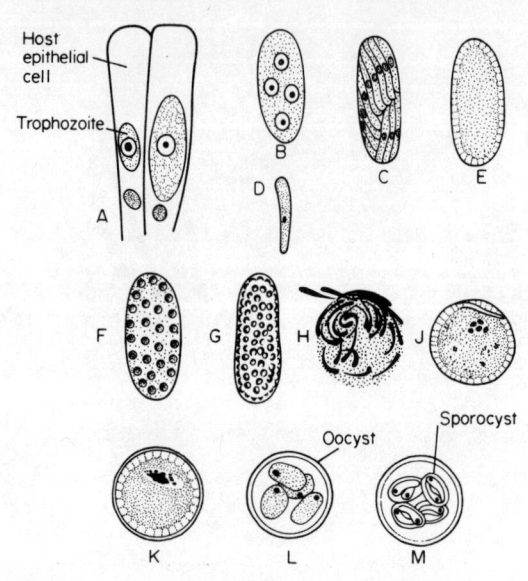

Fig. 2.5 Sporozoa. Life-cycle of a coccidian, *Eimeria* sp., from centipede intestine. A. Growing trophozoite in intestinal epithelial cells of host. B. and C. Schizogony. The trophozoite divides to form merozoites which re-infect host cells or develop into gametocytes. D. Merozoite. E. Gametocyte. F to H. Stages in formation of sperm from a male gametocyte. J. Fertilization of female gametocyte. K. Oocyst. L and M. Stages in sporogony, which produces 4 sporocysts each containing 2 sporozoites. The sporozoites cause new infection, developing into trophozoites within host epithelial cells (A—M after Hyman).

b Phototrophic protozoans possess cytoplasmic organelles called chromoplasts which may be green (chloroplasts where the photosynthetic pigment is chlorophyll), yellow, red or brown and in which photosynthesis occurs. They may have pyrenoids associated with them, where sugars are converted to storage products.

Of the heterotrophic protozoans, osmotrophs have no special organelles for food absorption.

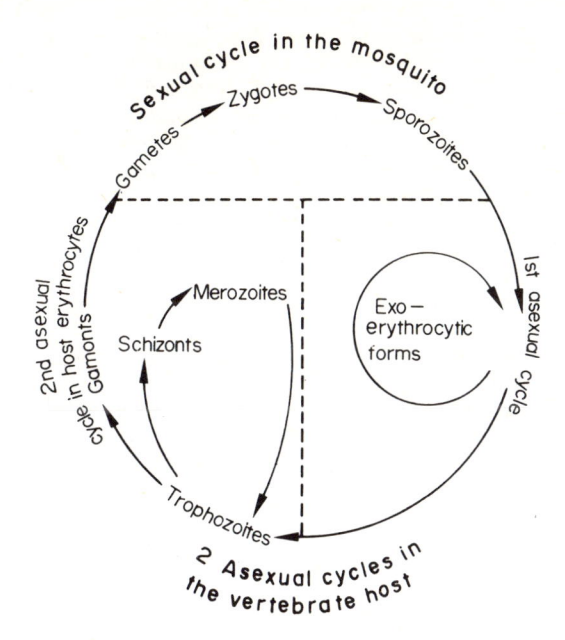

Fig. 2.6 Sporozoa. Scheme of the life-cycle of the Plasmodiidae (genus *Plasmodium*) (after Sergent).

Phagotrophs possess organelles and mechanisms for food capture and digestion. The former are very variable. Species which seek out and catch their prey or food either engulf it in an amoeboid way at any point on the surface (e.g. *Amoeba*) or in a particular region (many flagellates feed this way), or have a large, sometimes distensible cytopharynx and food enters the cytopharynx through the mouth or cytostome (ciliates). The so-called gullet of flagellates is not necessarily used in feeding. Where engulfment is amoeboid it is by means of pseudopodia which are variable in form in different groups. Filter-feeding and hunting may be combined, as in motile ciliates, where there is an extensive buccal ciliature which sweeps food into the mouth. Suctorians use tentacles for trapping and ingesting food; heliozoans use axopodia; radiolarians use axopodia and pseudopodia; foraminiferans and some testaceans use filopodia and rhizopodia. Trichocysts, of unknown function, are present in some species. Toxicysts may aid in food capture.

Digestion of solid food occurs in cytoplasmic food vacuoles. Where amoeboid engulfment occurs it may occur as circumvallation where particles are enclosed within a fluid medium, or by circumfluence where the pseudopodia are in close contact with the food particle and little or no fluid is taken up into the food vacuole. In species which feed through a gullet or cytopharynx food vacuoles are formed at its base, and these move off into the cytoplasm.

The food vacuoles undergo acid and alkaline phases during digestion, with enzymes being secreted into the vacuole. Absorbable materials pass into the cytoplasm.

Osmoregulation/excretion

a Many protozoans, particularly fresh-water species, possess one or more contractile vacuoles whose number, structure and position vary with the group or species, which may be fixed in position, and which may empty (as in ciliates) through a distinct pore in the pellicle.

The vacuoles are cavities which undergo cyclic processes of filling and emptying to the exterior. They may be permanent or impermanent structures. Two or more accessory vacuoles may unite or empty their contents into a large one which then empties or, as in ciliates, a complex system of canals may empty their contents into a vacuole.

In species with a definite body covering the ejection spot for solid wastes is called the anal spot or cytoproct.

b The contractile vacuole regulates the water content of the animal, eliminating excess water entering through the body surface or with food. Its involvement, if any, in excretion, has yet to be demonstrated.

Soluble waste products are removed across the general body surface. In naked protozoans solid waste such as undigested food in vacuoles is extruded through the body surface. In protozoans with a body covering it is removed through the cytoproct.

The main nitrogenous waste product is ammonia with urea and urates occurring less often.

Movement

a Locomotion may be by swimming, slow gliding, stepping (using cirri), rolling (using axopodia), or by peristaltic waves (euglenoids and some sporozoans). Some species are planktonic, others sessile. Sessile forms may have contractile stalks.

Phylum PROTOZOA

b   The three main types of locomotory organelles are pseudopodia, flagella and cilia. Contractile axostyles are found in some flagellates and undulating membranes (not of ciliary or flagellar type) in some sporozoans.

Pseudopodia are temporary protoplasmic extensions of the cell. If broad and lobose as in naked amoebae they are called lobopodia; if slender and tapered as in shelled amoebae, filopodia; if thread-like, branched and anastomosing as in foraminiferans, reticulopodia; if fine and needle-like with central axial rods as in heliozoans, they are called axopodia.

The flagellum is a thread-like structure arising from a basal body or kinetosome (blepharoplast). The basal body may itself act as a centriole. The flagellum has an ultrastructure similar to that of a cilium. It is made up of a number of smaller fibrils arranged in a characteristic way, enclosed by an extension of the plasma membrane. Flagella occur singly or in small numbers, except in hypermastigines where there are many.

Cilia usually occur in large numbers and are generally shorter than flagella. They may be separate or combined in compound structures such as cirri, membranellae or undulating membranes.

The ciliature of an organism may be separable into somatic (body) and oral ciliature. In ciliates the primitive condition would be to have uniform ciliation, with the cilia occurring in longitudinal rows, but in many species the cilia are reduced in number or restricted to certain areas of body surface.

Each cilium arises from a basal body or kinetosome from which a fibril or kinetodesma proceeds, joining a longitudinal bundle of similar fibrils from other kinetosomes of the same row. The kinetosomes and kinetodesmata together form the kinety.

Some sessile ciliates, e.g. *Vorticella*, *Stentor*, can undergo shortening. This is a function of the stalk in *Vorticella* and the whole body in *Stentor* and the responsible organelles are ectoplasmic myonemes in the longitudinal axis.

c   The slow gliding exhibited by the feeding stages of sporozoans is called gregarine movement. Its mechanism is unknown. Sporozoans may also move by body flexions.

Movement in sarcodines is accomplished by means of pseudopodia (e.g. reticulopodia, lobopodia) but amoeboid movement itself is found only in species with lobopodia or filopodia. Basically it is accomplished by cytoplasmic streaming and probably occurs in several variations in different species. In *Amoeba proteus* the ectoplasm and outer endoplasm form an outer semi-rigid tube of 'plasmagel' enclosing an inner more fluid zone of endoplasm, the 'plasmasol' (Fig. 2.7A). During locomotion gel is continuously converted to sol at the posterior end (uroid) of the amoeba and sol to gel at the anterior end (fountain zone). Although the sleeve of plasmagel does not itself move, the anterior end of the organism is continually being displaced forward. The motive force causing the plasmasol to advance in the plasmagel tube is not yet definitely known, whether a pull is exerted from the anterior end or whether pressure is exerted from the posterior end. When a sol is converted to a gel a decrease in volume occurs and possibly the plasmasol flows forward to fill the resulting space.

The flagellum propels the organism by a lashing or undulating motion (Fig. 2.7B). The former motion has a down- or effective stroke and a recovery stroke. Where undulation occurs the wave generally passes from base to tip of the flagellum. Where flagellar beat is oblique or where a spiral undulation passes along the flagellum, the organism usually rotates on its axis as it moves forward and may also gyrate. Forward, backward and lateral movements may occur. In some species (e.g. *Trypanosoma*) the flagellum may lie parallel to the body surface, connected to it by a thin, protoplasmic, undulating membrane.

The cilium effects locomotion by means of a lash or beat in one plane (Fig. 2.7C), with the effective stroke directed posteriorly and somewhat diagonally to the axis of the longitudinal row of cilia. Within each ciliary row the beat is sequential and a metachronal rhythm is set up, e.g. in *Paramecium* each cilium beats slightly later than the one in front, but simultaneously with its neighbours in adjacent rows.

The oblique beat of the cilia causes the organism to rotate on its longitudinal axis and to gyrate or follow a spiral course. The direction of movement may be reversed by reversal of the ciliary beat.

Co-ordination

a *i*   In ciliates the infraciliature is a complex system of fibres (kinetodesmata) underlying and connecting with the surface cilia. It *may* be responsible for co-

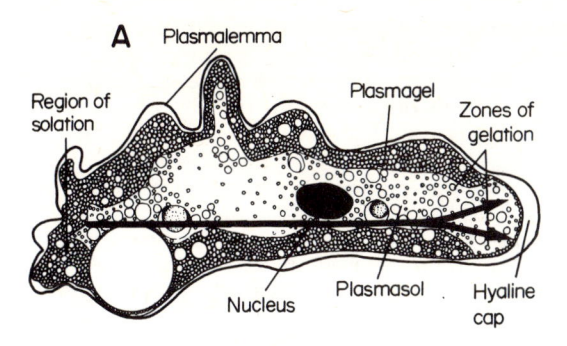

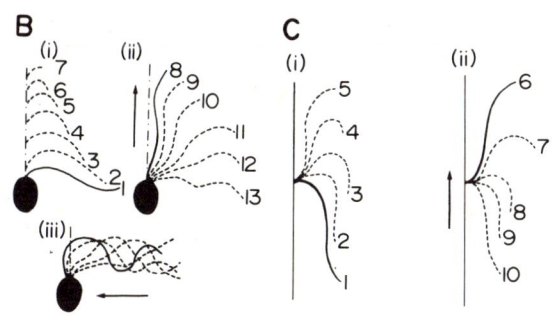

Fig. 2.7 Examples of protozoan movement. A. The amoeba, *Chaos diffluens*, showing regions of formation of plasmasol and plasmagel and the direction of endoplasmic currents (arrows) (after Mast). B. Flagellar movements of *Monas* (after Krijgsman). (i) Preparatory stroke. (ii) Effective stroke. (iii) Lateral movement. C. Ciliary action (after Verworn). (i) Preparatory stroke. (ii) Effective stroke. (Arrows in B and C show the direction of locomotion, numbers indicate successive positions of flagellum or cilium).

ordinating and controlling the ciliary beat, as may the rhizoplasts of the flagellar root system for controlling flagellar beat, but this has yet to be convincingly demonstrated. Ciliary co-ordination is usually considered to be achieved by visco-mechanical coupling between adjacent cilia.

Sensory organelles include light-sensitive eye-spots (in many phytoflagellates) and cilia and flagella which are sensitive to touch. Where sensory organelles are lacking the response to various stimuli such as light, chemicals, temperature, touch etc. is probably a general cytoplasmic response.

ii There are two types of behaviour in response to stimuli. In kineses the stimulus causes an increase in random movement. In taxes the response to the stimulus is directional (e.g. phototactic response in flagellates with eye-spots).

Respiration

a There are no special organelles.

b Respiratory exchange occurs across the body surface. Most protozoans show aerobic respiration.

Circulation/coelom

a There is nothing corresponding to a circulatory system as in metazoans.

b Internal transport of materials is probably due to diffusion but may be aided by endoplasmic streaming in some species.

Reproduction

a Gametes are produced by modification or division of single cells. They may be isogametes (identical in appearance) or anisogametes (differing in size and structure). Anisogametes range from those showing a slight difference in size to well-differentiated egg- and sperm-like gametes.

b *Sexual reproduction.* Parthenogenesis occurs in some species. Sexual reproduction occurs in all classes but not all class members. It is rare in flagellates.

i Where fusion of gametes occurs the process is called syngamy. In the flagellate *Chlamydomonas* syngamy occurs between isogametes of two different mating types. Meiosis or reduction division commonly occurs in the formation of the gametes but in some autotrophic flagellates and in some sporozoans is post-zygotic.

The formation of the zygote is frequently followed by division into two or more daughter organisms.

ii In ciliates the process of *conjugation* occurs. Gametes are not formed but two individuals (of opposite mating types but identical variety) adhere together, the intervening membranes break down and nuclei are exchanged. Each migrating nucleus fuses with a stationary one in the other individual to form a zygote nucleus.

Ciliates generally possess two types of nucleus — a large macronucleus (vegetative, responsible for general cellular functions) and one or more micronuclei (essential for sexual processes). During sexual processes genetic recombination and macronuclear regeneration occur.

In *Paramecium aurelia* for example, after the two conjugants mate, the macronucleus of each degenerates and the two micronuclei divide twice to form eight (meiosis having occurred at this stage). Of the eight micronuclei seven disappear, whilst the remaining one divides again. Of the two daughter nuclei one is stationary, the other migratory. After exchange of micronuclei and fusion, macronuclear regeneration from a micronucleus occurs in each conjugant.

Variants of conjugation are cytogamy where two ciliates come together but do not exchange nuclei, and autogamy where no mating occurs, the two micronuclei within a single cell fusing to form a zygote nucleus.

c *Asexual reproduction* occurs in all protozoans and in some may be the only mode. It occurs by fission.

i Binary fission produces two similar daughter cells. In flagellates fission is usually symmetrogenic, proceeding longitudinally between the rows of basal bodies. In ciliates it is homothetogenic, occurring transversely across the lines of basal bodies.

ii Budding produces one daughter cell much smaller than the other.

iii Multiple fission or schizogony produces a number of daughter cells. In most sporozoans sporogony (or spore formation) is a form of multiple fission which occurs after sexual reproduction.

Encystment

Cyst formation is a feature of the life-cycle of many protozoans, especially fresh-water species. The organism secretes a protective envelope around itself and becomes inactive.

Encystment may be a means of surviving adverse conditions such as drought and low temperatures. However encysted zygotes are also found, as are reproductive cysts within which processes such as fission or formation of gametes occur.

Classification

The classification of the Protozoa is problematical. The relationships between the so-called 'classes' of the phylum are not well understood. Additionally, with the exception of the ciliates, the classes may contain groups with superficial similarities whose relationships with one another are obscure. New classifications are being introduced at the present. That used here has the merit of being simple.

References

Sandon H. 1968. *Essays on Protozoology*, 2nd edition. Hutchinson, London.
Sleigh M. 1973. *The Biology of Protozoa*. Arnold, London.
Vickerman K. & Cox F.E.G. 1967. *The Protozoa* (Introductory Studies in Biology). John Murray, London.

3 Phylum MESOZOA

About fifty species are known, all minute.

Characteristics

1 Parasites of body cavities of various advanced invertebrates.
2 Metazoan, of simple organization.
3 Body of two cellular layers but lacking endoderm and mesogloea.
4 Body consists, during at least part of the life-cycle, of outer ciliated cells (the somatoderm) and inner reproductive cells.
5 Life-cycle complex.
6 Of uncertain affinities.

Larval form

Dicyemida. The oval larva is called the infusoriform dispersal larva and is found in the mature cephalopod host. It has two apical unciliated cells filled with high density material, and several large ciliated cells covering the surface (Fig. 3.1A). Its interior is called the urn. It escapes from the vermiform parent into the host urine and then into the sea. So-called stem vermiforms, seen in the early stages of cephalopod infection, are believed to be derived from some of the cells of the infusoriform.

Orthonectida. The larva resembles the infusoriform larva. After escaping from the entoparasitic parent, and after a migratory period, the larva infects a new host.

Metamorphosis

There is no metamorphosis.

Adult body form

Order *Dicyemida*. e.g. *Pseudicyema*. Common parasites of the kidneys of squids and octopus, the form generally observed is the vermiform (Fig. 3.2) or nematogen which is small, ciliated, worm-like, has a constant number of cells and an outer ciliated somatoderm enclosing one or more axial cells.

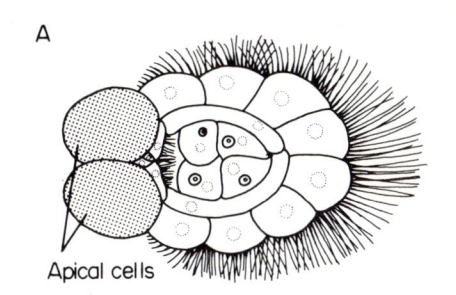

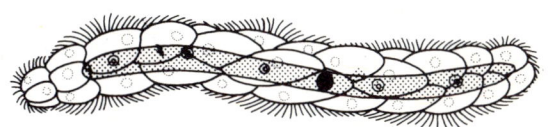

Fig. 3.1 Dicyemida. A. Infusoriform larva. B. Vermiform young (From *The Mesozoa* by Elliot A. Lapan and Harold J. Morowitz, December 1972. Copyright © 1972 by Scientific American, Inc. All rights reserved.)

Order *Orthonectida*. e.g. *Rhopalura* (Fig. 3.3). Rare internal parasites of various invertebrates, the asexual stage is a multinucleate amoeboid plasmodium which ultimately gives rise to males and females resembling the nematogen.

Recent thinking tends to consider orthonectids as a completely separate group.

There is no information on feeding, osmoregulation, excretion or respiration but presumably absorption of nutrients, osmoregulation, excretion and respiratory exchange occur across the outer layer of cells, or the limiting layer of the plasmodium.

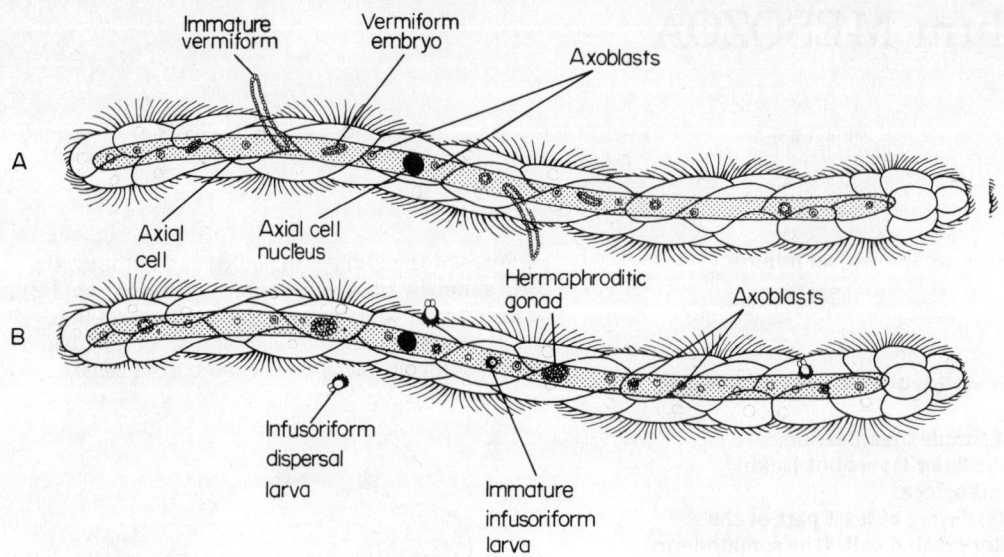

Fig. 3.2 Dicyemid vermiforms of similar morphological type. A. gives rise to vermiform young. B. produces infusoriform larvae. Axoblasts in A develop directly into vermiforms, in B they form hermaphroditic gonads, the fertilized eggs of which develop to infusoriforms (From *The Mesozoa* by Elliot A. Lapan and Harold J. Morowitz, December 1972. Copyright © 1972 by Scientific American, Inc. All rights reserved.)

Movement

Presumably swimming occurs in ciliated forms. Movement of the plasmodium stage may resemble that of amoebae.

Co-ordination

Specialized nervous structures are lacking.

Reproduction

a There are no special organs.

b *Sexual reproduction.* If the vermiform population in the host kidney reaches high density axoblast cells within the axial cell give rise, not to vermiform embryos but to a so-called hermaphroditic gonad. The eggs which it produces are fertilized by sperm, often from the same gonad,

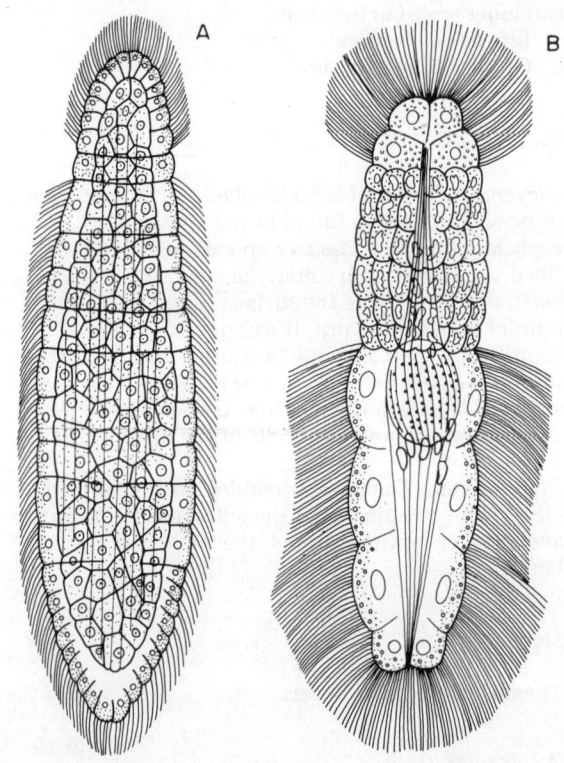

Fig. 3.3 The orthonectid *Rhopalura*. A. Female. B. Male (after Julin).

16   Phylum MESOZOA

and develop into infusoriform dispersal larvae which escape into the sea. It is now thought that an intermediate host is unnecessary for the transmission of infection.

In orthonectids the plasmodium ultimately gives rise to males and females resembling the nematogen, whose germ cells produce sperm and eggs. When they become sexually mature the males and females escape from the hosts and mate. Fertilization is internal, the zygotes developing within the female to larvae resembling the infusoriform larva.

c *Asexual reproduction.* In the dicyemid vermiform the axial cell contains a number of small reproductive cells or axoblasts. These may give rise to more vermiforms, which escape from the parent in an immature form and complete their development in the host kidney.

The plasmodium reproduces by plasmotomy, i.e. division producing multinucleate daughters.

Reference

Lapan E.A. & Morowitz H. 1972. The Mesozoa. *Scientific American*, **227**, No.6, 94–101.

4 Phylum PORIFERA

Several thousand species known, ranging in size from a few mm to one or two metres in height.

Characteristics

1 Marine except for one fresh-water family, the Spongillidae (Demospongiae).
2 Metazoan, with loosely aggregated cells, lacking organs.
3 Adults sessile.
4 Asymmetrical or radially symmetrical.
5 Body permeated with pores, canals and chambers through which the water current circulates.
6 Internal cavities are, at least partially, lined with flagellated choanocyte cells.
7 Possess an internal skeleton of spicules and/or organic fibres.
8 A nervous system is absent.
9 Possess high regenerative powers.

Larval form

After fertilization the embryo generally develops within the parental mesenchyme though some species liberate fertilized eggs. The flagellated embryo is liberated from the parent as a blastula or gastrula and swims for a time.

Metamorphosis

After swimming for some hours the larva attaches to the substrate, gastrulates where this has not already occurred, and develops into a small sponge.

Adult body form

The more primitive types of sponge are radially symmetrical and vase-shaped. More advanced types are asymmetrical, forming flat, rounded or branching structures which may be derived from the simpler *asconoid* form. The body surface is perforated by numerous openings, the smaller and more abundant incurrent pores or ostia and the larger excurrent pores or oscula.

The ostia and oscula are connected within the sponge by a system of canals and chambers. Fig. 4.1 is a diagram of the simplest type of organization, the asconoid type. The body wall has an outer layer of epithelial cells or pinacocytes and an inner layer (lining the central cavity or spongocoel) of choanocytes, the flagellated cells whose beating sets up the water current. Between the two is a

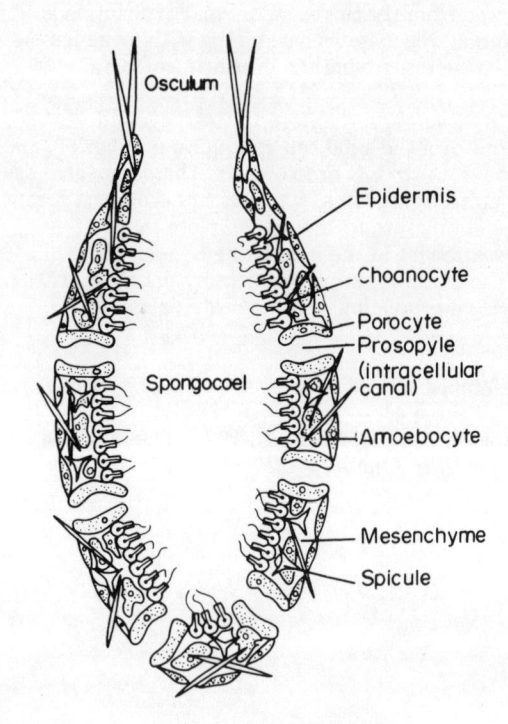

Fig. 4.1 Diagram of an asconoid sponge (after Hyman).

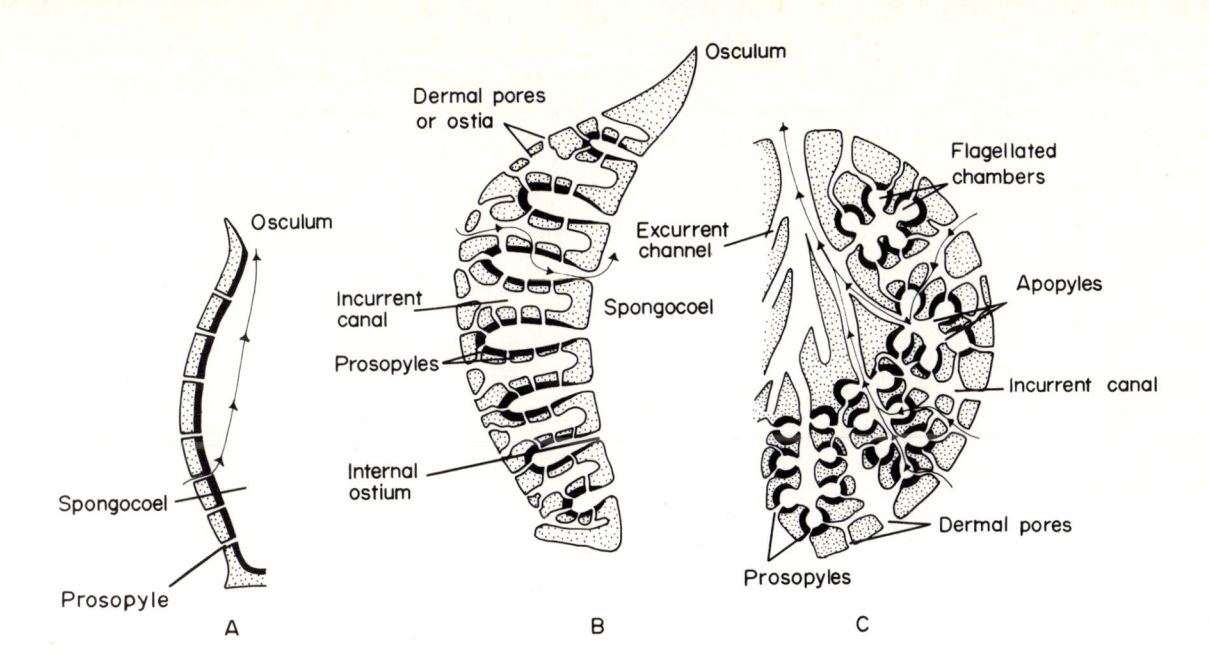

Fig. 4.2 Diagram of some types of sponge structure (after Hyman). A. Asconoid type. B. Syconoid, with cortex. C. Leuconoid type with eurypylous chambers. (Arrows show direction of the water current, stippled area the mesenchyme and solid black areas, the choanocyte layer.)

gelatinous mesenchyme in which the skeletal elements and various types of amoebocyte cell are found. At intervals in the body wall are porocyte cells through which an intracellular incurrent canal passes. Different degrees of outpushing of the body wall give rise to sponges with a more complex type of construction (Fig. 4.2).

Sponge classification is based largely on skeletal structures, which are separate crystalline spicules and/or organic fibres secreted by a type of amoebocyte called a scleroblast. Sponges are classified according to the chemical composition of the skeletal elements, the size of the spicules (megascleres or microscleres), and the number of axes or rays of the spicules (Fig. 4.3).

1 *Class* CALCAREA the calcareous sponges. Possess 1, 3 or 4-rayed spicules of calcium carbonate which are not differentiated into megascleres and microscleres; class includes sponges of simple asconoid form (e.g. *Leucosolenia*) and of syconoid or leuconoid form (e.g. *Sycon*).

2 *Class* HEXACTINELLIDA the glass sponges. e.g. *Euplectella*, *Hyalonema*. Possess siliceous spicules, 6-rayed (or some modification), separate or united in networks; lack a surface epithelium; choanocytes confined to finger-shaped chambers.

3 *Class* DEMOSPONGIAE e.g. *Plakina*, *Halichondria*. Possess 1 to 4-rayed siliceous spicules and/or horny (spongin) fibres (no skeleton in some primitive forms); form of construction leuconoid; class includes some fresh-water forms, e.g. *Spongilla*.

Feeding

a Sponges are filter-feeders.

b There is no digestive tract. Water entering the sponge passes through a series of openings of decreasing size, which act as sieves to filter out minute organisms and organic detritus on which it feeds. The openings are the ostia, prosopyles (Fig. 4.2)

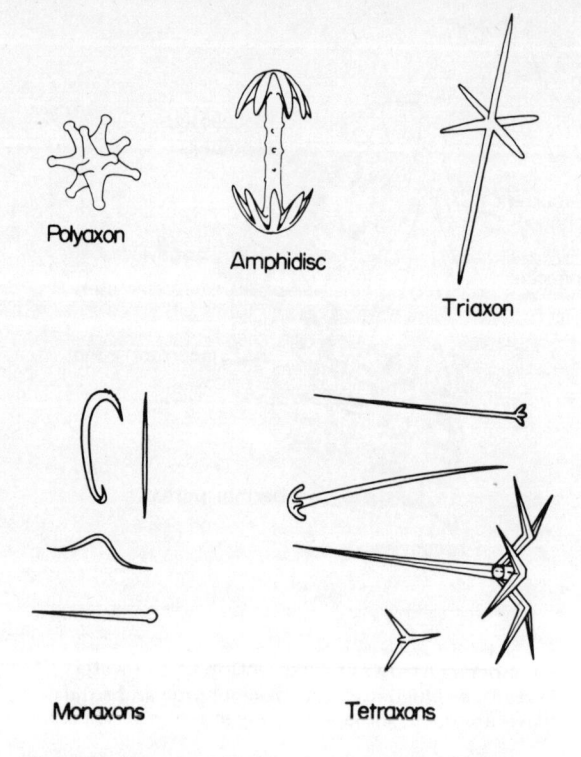

Fig. 4.3 Sponge spicules of various types (after Hyman and Meglitsch).

and spaces between the cytoplasmic tentacles of the collars of choanocyte cells. Ingestion is a function of choanocyte cells (which may then transfer the food to amoebocytes) and amoebocytes called archaeocytes. Large particles may be ingested by pinacocytes. Amoebocytes probably play the major role in digestion, which is intracellular, occurring in food vacuoles which undergo acid and alkaline phases as in protozoans.

Sponges probably also use dissolved nutrients. They store food reserves in a type of amoebocyte cell called a thesocyte.

Osmoregulation/excretion

a There are no special organs.

b In forms which have been studied appreciable amounts of ammonia are excreted.

Indigestible food wastes are expelled from thin-walled vacuoles of the archaeocytes, into an exhalent canal, or from the body surface.

Movement

a The adults are generally sessile.

b Where locomotion does occur it is by simple amoeboid spreading of the basal pinacocytes. Re-aggregation is a property of individual cells that re-establish colonial form after physical separation.

Co-ordination

a *i* There is no physiological evidence to demonstrate the presence of nerve cells in sponges. Some claims for their presence have been made on anatomical evidence.

ii In their responses to various types of stimulus individual sponge cells appear to act as independent effectors. Only slight conductive powers exist, e.g. strong tactile stimuli being transmitted only a few mm. By the slow contraction of myocytes (circularly arranged around pores or canals), desmacytes (bipolar cells with protoplasmic extensions ramifying throughout the inhalent canal system) and pinacocytes, the size of ostia and oscula can be altered, and the body surface area reduced.

b Hormones are unknown.

Respiration

There are no special organs. Respiratory exchange is a function of individual cells.

Circulation/coelom

a There are no special organs, and no coelom.

b Water enters the sponge through the ostia and leaves by the oscula. The current is driven by the unco-ordinated flagellar beating of the choanocytes. In the chambers of syconoid and leuconoid sponges, the lumen of the excurrent opening to the

Phylum PORIFERA

spongocoel, the apopyle, is much greater than that of all the incurrent openings or prosopyles and water enters the chambers at a greater velocity than it leaves. The collars of the choanocytes are so oriented as to direct the current towards the apopyle.

Reproduction

a   There are no special organs. The sex cells generally are derived from archaeocytes or choanocytes, and are usually found in the mesenchyme.

b   *Sexual reproduction.* Most sponges are hermaphrodites, with the ova and sperm being produced at different times to ensure cross-fertilization. The mature sperm, which enter the sponge in the water current, are often carried to the egg in a cell derived from a choanocyte which then fuses with the egg, freeing the sperm. Some species liberate fertilized eggs, others incubate the larvae.

c   *Asexual reproduction* occurs by:

i   budding,

ii   the formation of reduction bodies (a compact cell mass remaining after disintegration of the sponge body) under adverse conditions, which develop when conditions improve into young sponges,

iii   the formation of *gemmules* in fresh water and some marine species. The gemmules are another means of surviving adverse conditions, and germinate when conditions improve. They are composed of a mass of archaeocytes rich in food reserves, and in fresh-water species possess a hard coat. Marine gemmules develop external flagella at one pole and after swimming for a time attach by the opposite pole and develop into young sponges.

Regeneration

The Porifera have high regenerative powers and young sponges will develop from accidental fragmentation of the body. In some species the tips of branches regularly break off and regenerate.

Affinities

Unlike the Protozoa, the Porifera have a cellular construction but they differ from the Metazoa in lacking true tissues. They are currently considered to be metazoans occupying an intermediate position between the Protozoa and the Cnidaria but not directly related to other metazoan phyla. The Porifera and the Cnidaria may be descended from different ancestral groups of flagellates, or sponges may have diverged early from the cnidarian stem.

References

Hyman L.H. 1940. *The Invertebrates*, Vol. I. McGraw-Hill, New York.
Marshall A.J. & Williams W.D. 1972. *Textbook of Zoology, Invertebrates* (7th Edition, Parker & Haswell, Vol. I). Macmillan, London.

5 Phylum CNIDARIA

Contains about 9000 living species. Individual colonial polyps may be microscopic, most species macroscopic.

Characteristics

1 Majority marine, but with a few fresh-water species.
2 Metazoan, with tissues.
3 Phylum exhibits polymorphism. The two main structural types are the polyp and the medusa.
4 May be solitary or colonial.
5 Typical larval form is the planula.
6 Exhibit some form of radial symmetry.
7 Generally possess a ring of tentacles around the oral end.
8 The single body cavity is the coelenteron.
9 The body wall is three-layered.
10 Possess nematocysts, special cell organelles used for offence and defence, located in cnidoblast cells of epidermis and gastrodermis.
11 Possess undifferentiated interstitial cells, which give rise to sex cells and nematocysts, and are involved in regenerative and reproductive processes.
12 Nervous system a network and not centralized.
13 Hermaphroditic or dioecious.

Larval form

The fertilized egg, which may or may not be brooded, generally develops into a ciliated stereogastrula, the planula larva.

Metamorphosis

Generally, after swimming for a few hours to many days the planula attaches and develops into a polyp or polypoid form, which in colonial species subsequently gives rise to the colony.
In some hydroids the planula remains in the

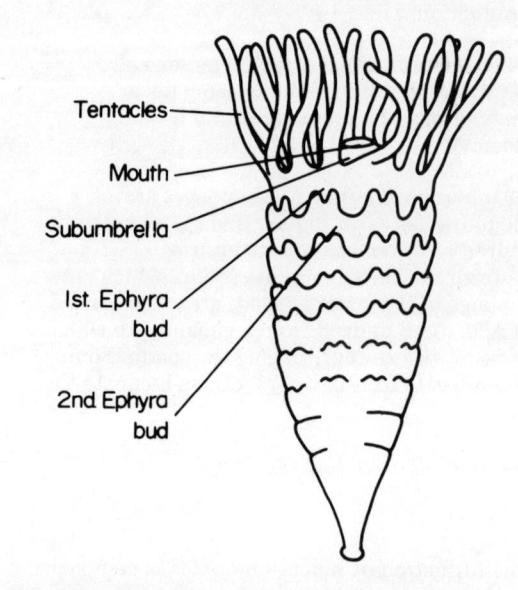

Fig. 5.1 A scyphistoma of *Aurelia*, budding ephyrae (after Bullough).

gonophore, developing into a tentaculate actinula larva which is liberated and creeps about. After attachment it develops into a polyp. In many hydrozoans with no polypoid phase the planula develops into an actinula and then a medusa.
In most scyphozoans, after attachment, the planula develops into a polypoid *scyphistoma*, with a stalked trumpet-shaped body (Fig. 5.1). At maturity, the scyphistoma produces a free-swimming medusa stage, the *ephyra* larva, by transverse fission or strobilization. Ephyrae may be produced singly or several at a time, and develop to adult medusae.
In zoantharian anthozoans, the planula does not attach but develops into an anemone-like Edwardsia larva, then the Halcampoides larva. After attachment, tentacles develop and the adult polyp form is attained.

Adult body form

Cnidarians exhibit some form of radial symmetry and generally have one or more whorls of tentacles encircling the oral end. The body wall has 3 layers, an outer epidermis, the mesogloea and an inner gastrodermis. The gastrodermis lines the single body cavity, the gastrovascular cavity or coelenteron.

The phylum is noted for polymorphism (see later), or variety of body form shown by a single species. The two main morphological variations are the polyp and the medusa. The polyp is generally sessile, and the medusa is generally free-swimming and the sexually-reproducing form.

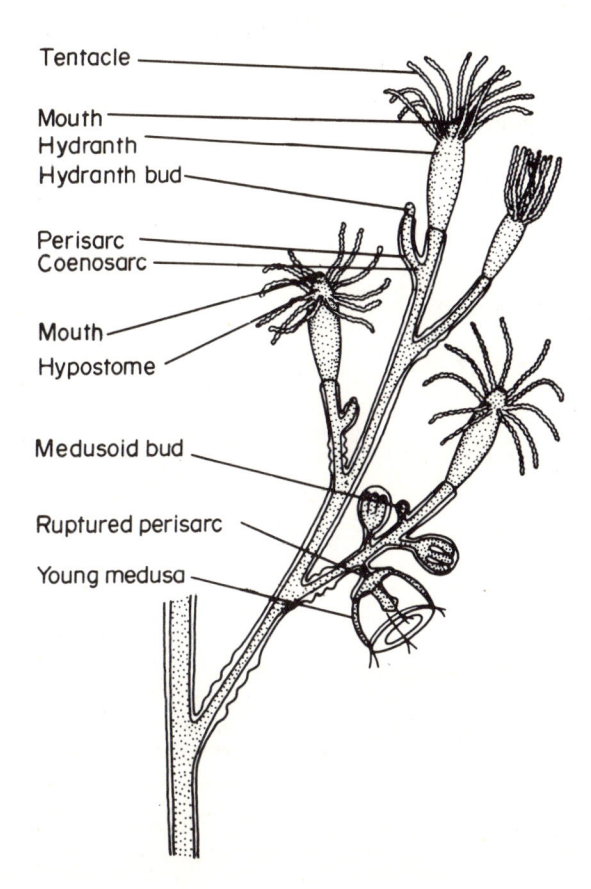

Fig. 5.2 Part of a colony of *Bougainvillea* (Hydrozoa) (after Bullough).

Labels: Tentacle; Mouth; Hydranth; Hydranth bud; Perisarc; Coenosarc; Mouth; Hypostome; Medusoid bud; Ruptured perisarc; Young medusa.

The polyp has a cylindrical form and is attached basally by a pedal disc or root-like stolons. In the Hydrozoa the cylinder or column bears an oral cone (or hypostome) with a terminal mouth and a ring of tentacles encircling the base of the cone (Fig. 5.2), and in the Anthozoa an oral disc, with elongated mouth, and tentacles encircling the disc.

The medusa is shaped like a dome, bell or umbrella with a convex aboral surface, the exumbrella, and a concave oral surface, the subumbrella. The bell bears tentacles around its margin. From the centre of the subumbrella projects the manubrium, a tube lined with endoderm. Its free end bears the mouth and its other end leads into the stomach, occupying the central region of the bell. The stomach in turn leads into radial gastrodermal canals which connect with a ring canal in the margin of the bell. The bulk of the medusan bell is composed of mesogloea.

The skeleton may be hydrostatic, or an exoskeleton secreted on the external surface, or an endoskeleton formed in the mesogloea as separate elements or a continuous mass. In medusae the mesogloea of the bell provides support.

1 *Class* **HYDROZOA** the hydroids and velum-bearing medusae, e.g. *Hydractinia, Eutonina*. Radial symmetry tetramerous or polymerous; solitary or colonial; life-cycle may include both polypoid and medusoid forms or may lack either; no stomodaeum, gastric tentacles or septa in coelenteron; non-cellular mesogloea; sex cells mature in the epidermis; oral end of polyp elongated into a hydranth (stomach region, oral cone and tentacles); medusae generally have a fold or velum projecting horizontally inwards from the bell margin (Fig. 5.3A). In colonial forms the polyps are in continuity through their body layers and their common gastrovascular cavity (Fig. 5.2). Members of the colony may be differentiated to perform different functions.

The epidermis commonly secretes a horny exoskeleton, the periderm (or perisarc). This is calcareous and massive in the Milleporina and Stylasterina.

Examples:
Hydra — solitary polyp, no medusoid stage;
Obelia — colonial, with polypoid and medusoid forms;
Geryonia — medusa, no polypoid stage;
Velella — colonial, pelagic, medusoid and polypoid individuals.

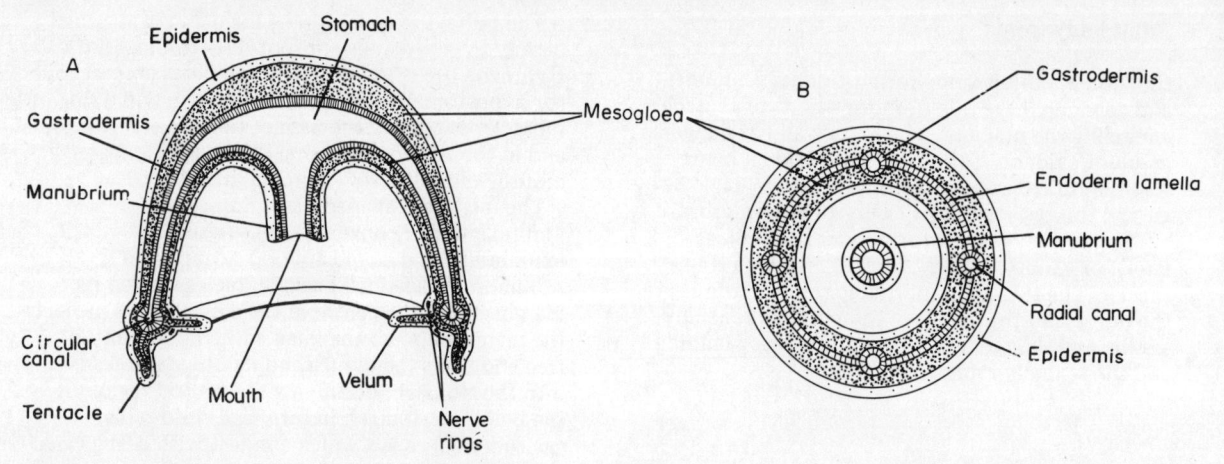

Fig. 5.3 A. Diagrammatic hemisection of a hydrozoan medusa, passing through the manubrium. B. Transverse section of the same (A and B after Parker).

2 *Class* SCYPHOZOA   the jellyfish, medusae lacking a velum, e.g. *Cassiopeia, Haliclystus*. Radial symmetry tetramerous; where not free-swimming then attached by an aboral stalk (Fig. 5.4A); coelenteron lacks a stomodaeum but has gastric filaments and may be divided by septa into four pockets; mesogloea cellular; gonads endodermal (gastrodermal); life-cycle generally includes a polypoid stage, the scyphistoma, which develops directly to the adult or produces medusae by transverse fission.

A skeleton is lacking, but the mesogloea of the bell gives support.

Examples:
Lucernaria — stalked;
Aurelia (Fig. 5.4B), *Cyanea* — free-swimming.

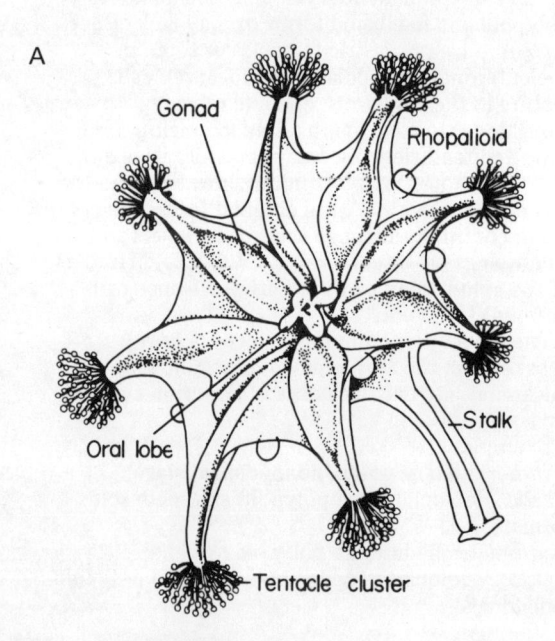

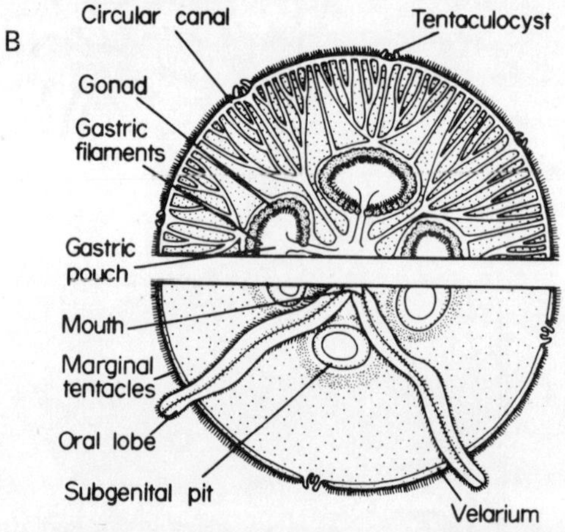

Fig. 5.4 A. A stalked scyphozoan, *Haliclystus* (after Hyman). B. Upper half of diagram shows the gastrovascular and reproductive systems of *Aurelia* (a scyphozoan medusa) as seen through the exumbrella; lower half of diagram shows the subumbrellar surface (after Bullough).

24   Phylum CNIDARIA

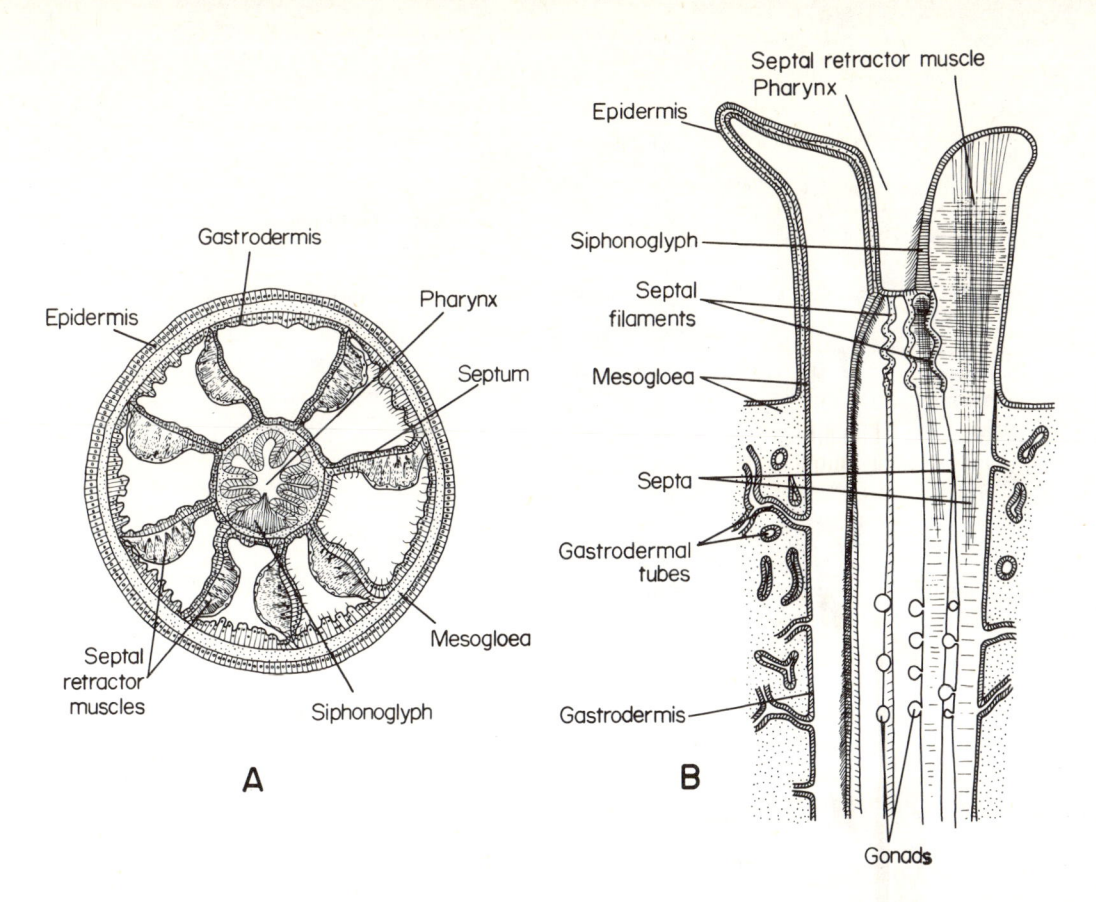

Fig. 5.5 A. Transverse section through the pharynx region of *Alcyonium*, showing the siphonoglyph and arrangement of the septa (after Hickson).

B. Diagrammatic longitudinal section of an alcyonacean polyp (after Hyman).

3 *Class* ANTHOZOA the sea anemones and corals, e.g. *Eunicella*, *Actinia*. No medusoid stage; symmetry biradial; richly-cellular mesogloea; oral end expanded to an oral disc; inturning body wall at the mouth forms a stomodaeum (pharynx); coelenteron divided by septa with filaments (bearing nematocysts) on their free edges (Fig. 5.5A); gonads endodermal; solitary (Fig. 5.6) or colonial — colony members communicate by gastrodermal tubes (Fig. 5.5B).

The structural skeleton, where present, may be of calcium carbonate, secreted by the epidermis as in stony or madreporarian corals (Fig. 5.7) or may be an endoskeleton formed by mesogloeal cells, horny or calcareous, of spicules or some other unit or not. The spicules may or may not be united.
  Examples:
Tubipora — colonial form, endoskeleton of fused spicules;
Alcyonium — colonial, endoskeleton of separate spicules;
Metridium, *Tealia* — solitary, no secreted skeleton;
Porites — colonial, with exoskeleton;
Fungia — solitary, with exoskeleton.

Feeding

a Cnidarians are generally carnivorous and either trap prey or are suspension feeders.

Phylum CNIDARIA

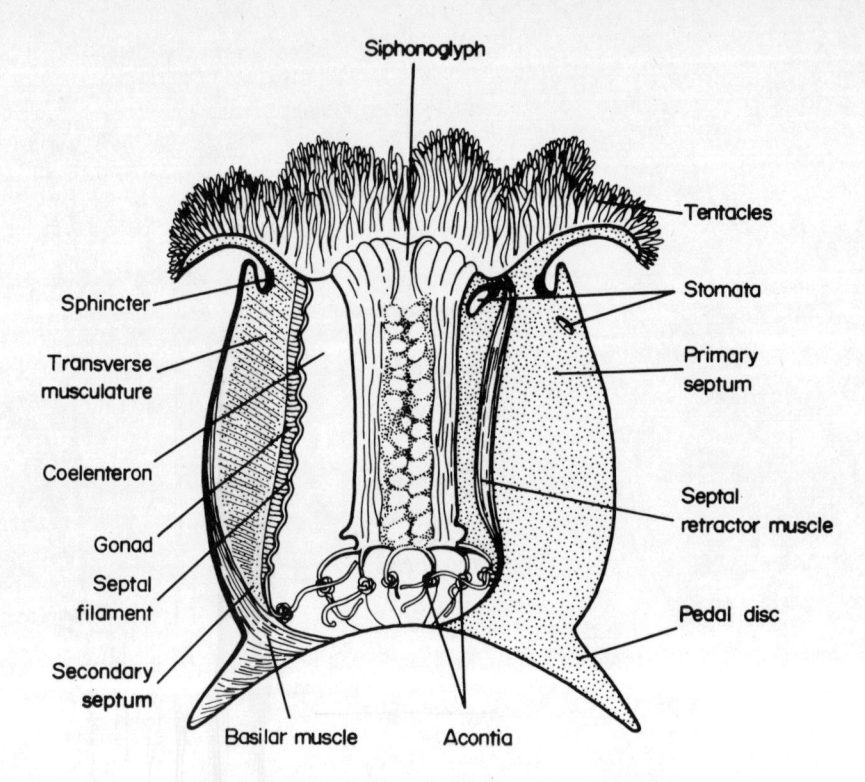

Fig. 5.6 Diagrammatic longitudinal section of an anemone (after Hyman).

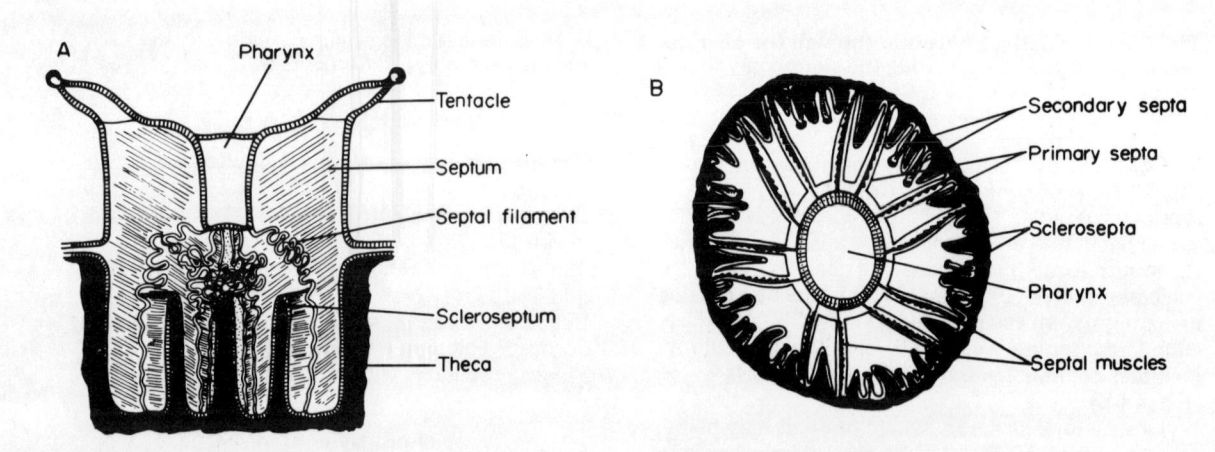

Fig. 5.7 A. Diagrammatic longitudinal section through a polyp of a colonial stony coral in its theca (after Hyman). B. Transverse section through the pharyngeal region of a stony coral polyp (after Duerden).

26 Phylum CNIDARIA

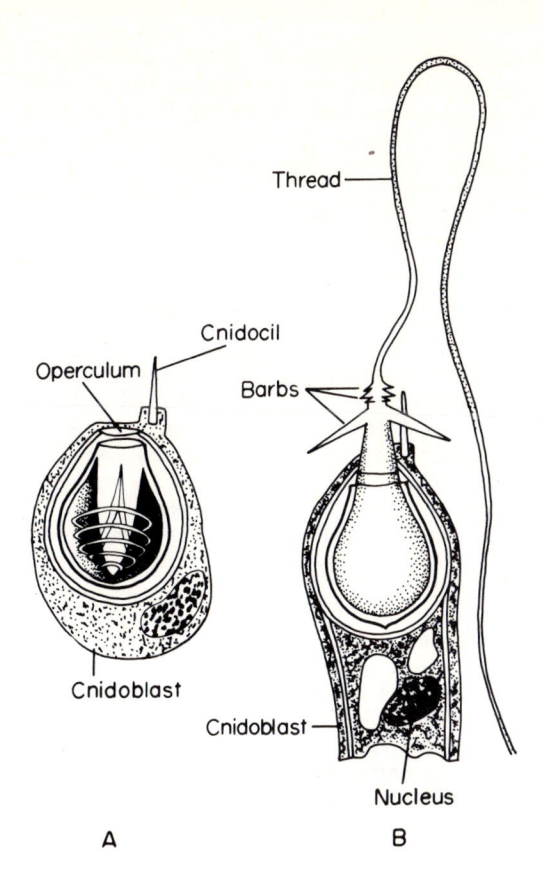

Fig. 5.8 A. Undischarged nematocyst of *Hydra*. B. Discharged nematocyst of *Hydra* (A and B after Schneider).

b The tentacles of cnidarians are used in prey capture. The tentacles and extensions of the manubrium such as oral arms, lobes and tentacles are well supplied with nematocysts which discharge when stimulated and aid in holding and subduing prey (Fig. 5.8).

The digestive system is the coelenteron. In polypoid types it may be simple as in the Hydrozoa, or partitioned by septa as in the Anthozoa. In medusae it consists of the manubrium and a central stomach connected by radial canals to a ring canal within the bell margin. In the Scyphozoa the stomach is divided by septa which bear gastric filaments containing nematocysts and gland cells. In the Anthozoa the free septal edges bear septal filaments with gland cells and nematocysts.

The tentacular and manubrial nematocyst threads entangle and paralyse prey which is then transferred to the mouth. In medusae prey is captured by extensible tentacles as the animal sinks slowly downwards in locomotion. Many scyphozoans and smaller anthozoans are ciliary suspension feeders and food particles are trapped in a mucous film or strands, e.g. on the column, tentacles, or subumbrellar surface, and are transferred to the mouth by ciliary beat.

Digestion begins extracellularly with gastrodermal gland cells secreting a proteolytic enzyme into the coelenteron. The food is reduced to a coarse suspension, flagellar currents and body movements probably aiding mixing, and both fluid and particles are then phagocytised into food vacuoles of gastrodermal cells. Intracellular digestion proceeds with acid and alkaline phases in the food vacuole, and is followed by absorption.

In colonial hydroids extracellular digestion occurs within the nutritive polyps and the resulting brie then passes into the common gastrovascular cavity for intracellular digestion.

Osmoregulation/excretion

a There are no special organs.

b Nitrogenous waste is mainly ammonia which diffuses out through the general body surface. Indigestible food materials are ejected through the mouth.

Movement

a Movement in polypoid forms is generally slow and restricted to tentacular and column movements, though occasionally they may change location, e.g. by floating, somersaulting (*Hydra*), slow gliding on the pedal disc (some anthozoans), swimming (*Stomphia*), or walking on the tentacles. A few anemones swim by means of tentacular or column movements. Siphonophores use jet propulsion.

Free-living medusoid forms float or swim.

b There are no special organs for locomotion.

c The simplest cnidarian muscular system is found in hydroid polyps. The epidermal epithelio-muscular

Phylum CNIDARIA

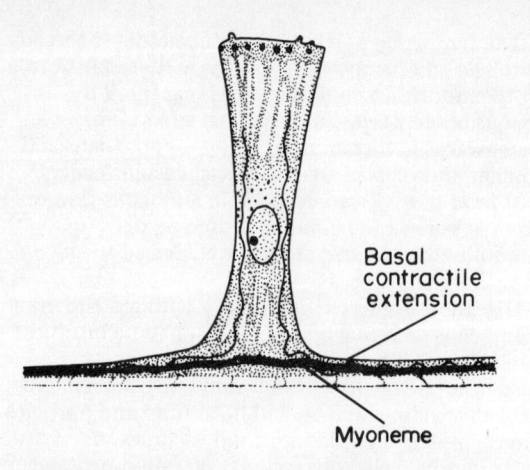

Fig. 5.9 An epidermal epithelio-muscular cell of a hydra (after Gelei).

cells (Fig. 5.9) and the gastrodermal nutritive-muscle cells have basal contractile extensions adjacent to the mesogloea. These extensions contain contractile fibres or myonemes, which in the epidermis are oriented parallel to the body axis and in the gastrodermis are oriented circularly. Tentacular and column movements are due to the actions of these two cylinders of contractile fibres on the hydrostatic skeleton.

Medusae possess a mainly epidermal musculature, with longitudinal fibres in the manubrium and tentacles and radial fibres in the subumbrella. Such fibres are smooth and extensions of epidermal cells. The main musculature of medusae, responsible for swimming movements, is composed of extra epidermal muscle cells, lying beneath the epidermis and running circularly in the velum (where present) and the subumbrella. These muscles are striated. Contraction of these muscles and the radial fibres of the subumbrella, causes rhythmic pulsations of the bell, which force water out from the subumbrellar space. When contractions cease the medusa drifts downwards.

In scyphozoan medusae the circular subumbrellar muscle is called the coronal muscle, which apparently resembles vertebrate muscle in its properties.

In the Anthozoa the epidermal system of fibres is reduced, generally to fibres in the tentacles and oral disc, and the main muscles are gastrodermal. The usual cylindrical layer of circular muscle is present and may form a sphincter, often lying in mesogloea, where the column joins the oral disc, enabling the latter to be covered when the tentacles are withdrawn. In addition to this muscle cylinder, longitudinal and transverse muscle bands are located in the coelenteric septa.

Co-ordination

a *i* The nervous system is of the nerve net type, generally synaptic, and composed of non-polarized multipolar and bipolar neurones, situated at the bases of the epidermis and gastrodermis. There is some concentration of cells and fibres in certain groups. Various types of sensory cell are present, including probably chemoreceptors and mechanoreceptors scattered in the epidermis and more sparsely in the gastrodermis. They may be concentrated on the tentacles and near the mouth. In some groups sensory cells may be concentrated in special organs.

In hydrozoan polyps and anthozoans there is a typical nerve net with little concentration of elements except perhaps round the mouth. Special sense organs are lacking.

Medusae generally have a more organized type of nervous system, reflecting their more active way of life. There is a sub-epidermal plexus in the manubrium and tentacles and between the two epithelio-muscular layers of the subumbrella. This may be more concentrated along the radial canals. In hydromedusae the nerve plexus connects with a nerve ring of neurone processes and nerve cells at the junction of the bell and velum. The ring is sub-divided, the upper part innervating the tentacles, ocelli, sensory tracts, and marginal patches of sensory epithelium, the lower innervating the balance organs and muscles of the subumbrella and velum. The main sense organs are light-sensitive ocelli on the tentacle bases or bell margin, and organs of balance, the statocysts, borne on the bell margin. Sensory tracts overlie the nerve rings.

Scyphomedusae have a similar nervous system to hydromedusae but, with the exception of the Cubomedusae, lack nerve rings. Marginal sense organs called rhopalia are present, in which are located statocysts, patches of sensory epithelium, and in some species ocelli. Closely associated with the rhopalia are concentrations of nerve cells called rhopalial ganglia.

ii Cnidarian nerve nets exhibit diffuse conduction and nervous excitation spreads in many directions from the point of stimulus. The amplitude of the muscle response is dependent upon the frequency of stimulation (fast and slow muscles) and on neuromuscular facilitation (fast muscles). Restricted spread of nervous pulses may be related to the requirements of interneural facilitation.

The nervous system may be differentiated into two or more types of net, separately connected to effectors and interacting only in limited regions. Commonly there is a fast, through-conducting net which mediates a specific response such as the swimming pulsation of medusae or polyp withdrawal, and a slow, diffuse, non-through-conducting net which mediates variable local movements. In hydromedusae the lower nerve ring is the centre of the rhythmic pulsations and contains the pacemakers. In scyphomedusae, with the exception of Cubomedusae, the initiation and control of pulsation are functions of the rhopalial ganglia.

Generally synapses are unpolarized. Many normally transmit impulses in a 1-to-1 fashion but others require facilitation by temporal or spatial summation of impulses. The nervous impulses of the cnidarian nerve net are of the same all-or-none type as in higher animals. Neuroid (non-nervous) conduction between epithelial cells has been demonstrated in some free-living Hydrozoa; the neuroid systems control contraction of certain muscles.

b Neurosecretory cells are reported present in the subhypostomal region of *Hydra*. Active substances stimulate growth and inhibit reproduction. A pheromone inhibiting budding is also elaborated.

Respiration

a There are no special organs.

b Gaseous exchange may be assisted by the circulation of coelenteric contents, due to body movement and ciliary or flagellar beating. For example, flagellar currents are set up in the canal system of some medusae and in many anthozoans water currents enter the coelenteron through the ciliated grooves or siphonoglyphs of the mouth.

Circulation/coelom

a There is no coelom and no haemocoel.

b The gastrovascular cavity serves for both digestion and absorption.

Reproduction

a There are both dioecious and hermaphroditic species.

The gametes of Hydrozoa originate from epidermal or gastrodermal interstitial cells (sometimes from epidermal or gastrodermal cells) and are typical ova and spermatozoa. The gonads are located at a characteristic site, e.g. on the column in *Hydra*, on the radial canals or manubrium of medusoid forms.

The gametes of Scyphozoa arise from gastrodermal interstitial cells and the gonads are formed on each side of the gastric septa or on the floors of the gastric pouches. In the Anthozoa the gametes are formed from gastrodermal interstitial cells and the gonads are located in the coelenteric septa.

b *Sexual reproduction.* Fertilization may occur *in situ* or in sea water. The fertilized egg may be shed into sea water or may develop *in situ* to a late stage.

c *Asexual reproduction* is common. It may occur by:

i budding, e.g. in *Hydra*, in colonial forms, in scyphozoan scyphistomae, in hydrozoan Anthomedusae which produce more medusae;

ii production of frustules (non-ciliated planula-like bodies which develop into polyps), e.g. in many hydrozoans;

iii pedal laceration, e.g. in sea anemones;

iv transverse fission, e.g. in the production of ephyrae by scyphistomae;

v longitudinal fission, e.g. in many sea anemones.

Regeneration

Cnidarians possess high regenerative powers.

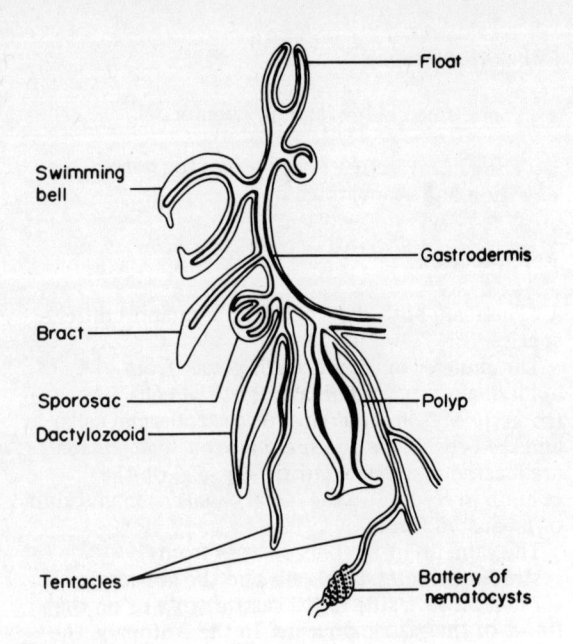

Fig. 5.10 Diagram of a siphonophore (after Claus).

Nematocysts

A nematocyst is a cell organelle which occurs within a cnidoblast cell (Fig. 5.8). It consists of a sac drawn out into a thread whose tip may be open or closed. The thread normally lies coiled within the sac and at the junction of the two is a lid or operculum. A stimulus to the cnidocil, on the outer surface of the cell, causes the nematocyst thread to be everted, and depending on type, to adhere to, coil round, or penetrate and paralyse prey.

Polymorphism

In many cnidarians the life-cycle contains two morphologically dissimilar individuals, the polyp and the medusa. In colonial species each of these types may occur in a number of different morphological forms, specialized to perform a particular function.

The main types of modified polyp are:
the gastrozooid — feeding polyp,
the gonozooid — reproductive polyp,
the dactylozooid — protective polyp,
or tentaculozooid
} found in hydrozoan colonies

the autozooid — feeding and reproductive polyp,
the siphonozooid — current-producing polyp.
} found in some anthozoan colonies

A colony may also bear medusoid forms in different stages of formation or degeneration, which may or may not be freed. Medusae may become modified as swimming bells, floats, protective bracts or phyllozooids, or gonophores which serve only for reproduction.

The order Siphonophora of the Hydrozoa shows the greatest degree of polymorphism of colonial forms (Fig. 5.10).

Reference

Hyman L.H. 1951. *The Invertebrates*, Vol. I. McGraw-Hill, New York.

6 Phylum CTENOPHORA

About 90 species known. Ovoid forms measure up to about 5 cm, flattened forms may be up to 1 metre or more in length.

Characteristics

1 Marine, most free-swimming, but a few creeping or sessile forms.
2 The larva (cydippid) is free-swimming.
3 Body spherical to ovoid, or flattened.
4 Most orders tentaculate.
5 Tentacles, where present, bear adhesive colloblast cells used in prey capture.
6 Transparent, gelatinous and luminescent.
7 Possess 8 radially-arranged ciliated bands or comb rows.
8 Locomotion by ciliary beat of the comb rows.
9 Digestive system of branched canals.
10 Hermaphroditic.

Larval form

The free-swimming cydippid larva, closely resembling the spherical to ovoid ctenophores, is of general occurrence throughout the group.

Metamorphosis

The cydippid larva necessarily undergoes a more extensive change in shape when attaining the body form of flattened or elongate species, than it does in spherical or ovoid species.

Adult body form

The more primitive ctenophores are those with the spherical to ovoid body form. There is a mouth at one pole (Fig. 6.1A) and an apical sensory organ at the other. The body bears eight ciliated bands or comb rows, extending from the apical organ almost to the mouth. In the aboral hemisphere a pair of tentacles proceeds from sheaths or pouches.

The more primitive ctenophores are included in

1 *Class* TENTACULATA e.g. *Mertensia, Pleurobrachia*. Representative orders may have a moderately compressed body and unsheathed tentacles (e.g. *Mnemiopsis*); a ribbon-shaped body, reduced tentacles and comb rows (e.g. *Cestum, Velamen*); or a body much flattened along the polar axis, comb rows reduced or absent in the adult and a creeping habit (e.g. *Ctenoplana, Coeloplana*).

The second class of ctenophores is

2 *Class* NUDA lacking tentacles, e.g. *Beroë*, with a conical body and large pharynx. There is no skeleton. Body form is maintained by a thick gelatinous layer, homologous to the coelenterate mesogloea, which is scattered with muscle cells arranged in an anastomosing network, amoebocytes, connective tissue fibres and probably nerve cells.

Feeding

a Ctenophores are carnivorous on plankton.

b The contractile tentacles, which lack nematocysts (except in the single species *Euchlora rubra*), bear adhesive colloblast cells in the epidermis (Fig. 6.1B). These are used in prey capture. Food caught on the tentacles is wiped off onto the mouth. Non-tentaculate forms probably prey on weaker and smaller organisms.

The digestive system consists of a series of branched canals. Digestion begins extracellularly in the pharynx and after passage of the broth to the stomach and canal system, is completed intracellularly.

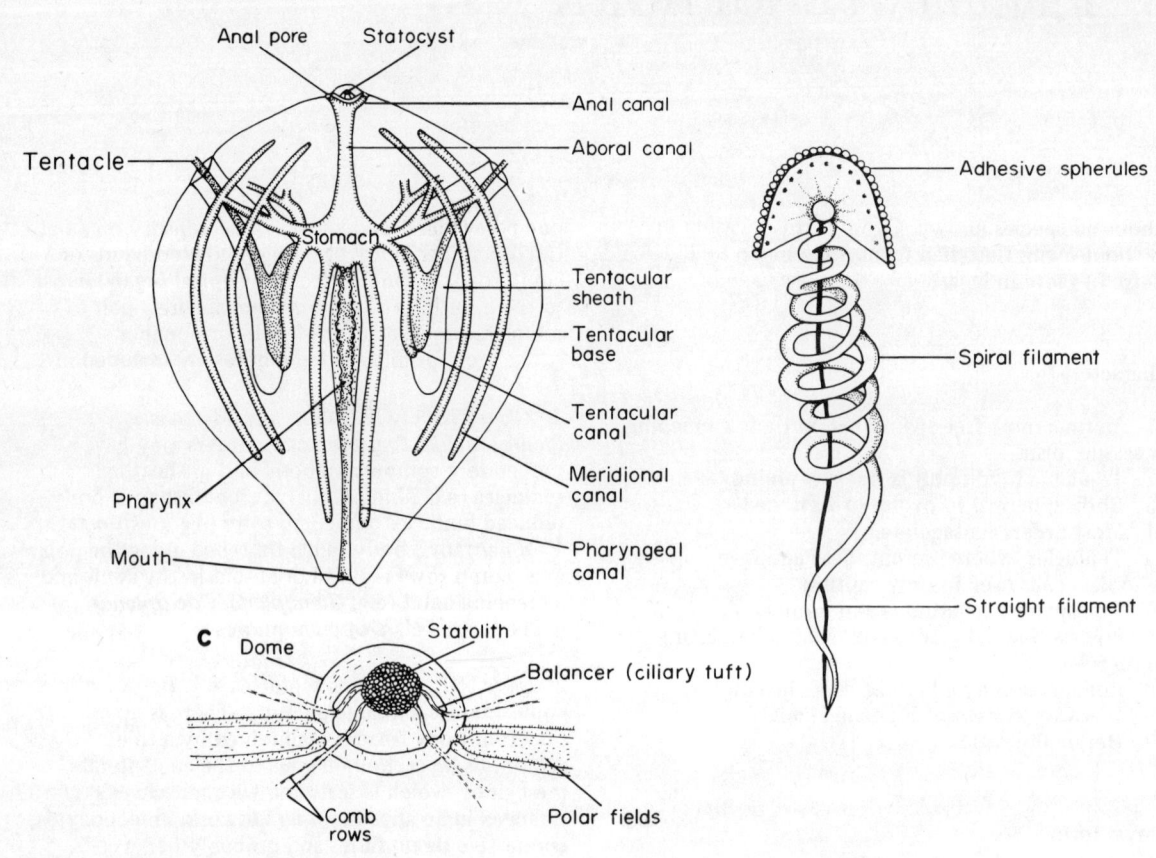

Fig. 6.1 A. Digestive system of a tentaculate cydippid ctenophore (after Hyman). B. Colloblast (after Komai). C. Arrangement of statocyst and comb plate cilia (after Chun).

Osmoregulation/excretion

a Groups of cells called cell rosettes are found in the digestive canals, which may play a part in osmoregulation or excretion. They consist of a circle of ciliated cells guarding an opening between the lumen of the canal and the mesogloea.

b Possibly waste or superfluous fluid may pass from the mesogloea through the openings and into the canals. Most nitrogen is excreted as ammonia, presumably from the coelenteric and general body surface. Indigestible wastes are removed through the mouth and anal pores.

Movement

a Depending on species, ctenophores move by slow swimming or creeping.

b In swimming forms the beating ciliary plates of the comb rows are responsible for movement. Normally waves of ciliary movement begin at the aboral ends of the rows and pass orally, but the beat may be temporarily reversed if necessary. The animal normally swims mouth upwards but may reverse its position, and can right itself if tilted. In forms such as *Velamen*, with a ribbon-like body, muscular undulations also occur during swimming.

Phylum CTENOPHORA

Such species have a well-developed muscle system. Creeping using a flattened ventral region occurs in *Tjalfiella, Vallicula*.

Co-ordination

a *i* The epidermal nerve plexus is concentrated as a ring surrounding the mouth, and at the bases of the comb rows, forming the radial nerves (not true nerves but condensations of the nerve net). Sensory cells are scattered in the epidermis, with some concentration near the mouth.

The sensory apical organ is a statocyst or balancer organ, used in maintaining normal orientation. The statolith of the organ rests on four tufts of balancer cilia (Fig. 6.1C). From each tuft a ciliated groove proceeds which forks, each fork running to a comb row and extending through it.

ii The nervous system and apical organ control the synchrony and co-ordination of the ciliary waves during swimming and the righting reflex. If the animal is tilted differential pressure is exerted on the tufts of balancer cilia, and signals are transmitted via the ciliated grooves to the appropriate comb rows, causing a change in the rate of beating. The ctenophore thus rights itself.

b Hormones are unknown.

Respiration

There are no special respiratory structures.

Circulation/coelom

a The gastrovascular cavity combines both circulatory and digestive functions. There is no blood system.

b The stomach and digestive canals are lined on the side nearest the polar axis with ciliated cells which probably provide the circulatory current.

Reproduction

a The gonads form two bands (one an ovary, the other a testis) located in the thickened wall of each meridional canal.

b Eggs and sperm are generally shed to the outside through the mouth, fertilization occurring in seawater. The exceptions are the few species (e.g. *Coeloplana*) which brood their eggs.

c A form of *asexual reproduction* may occur in some creeping species (e.g. *Ctenoplana*). Small fragments shed during locomotion, develop into complete ctenophores.

Embryology and development

Cleavage of the egg is determinate and there is mosaic development, with parts of the adult being mapped out during cleavage stages. However the possibility exists of a later reduction in determinative influence since adult ctenophores have considerable regenerative powers.

References

Horridge G.A. 1965. *American Zoologist* 5, 357–375
Hyman L.H. 1940. *The Invertebrates* Vol. I. McGraw-Hill, New York.

7 Phylum PLATYHELMINTHES

Several thousand species described ranging in length from a few mm to several metres.

Characteristics

1 Parasitic or free-living metazoan flatworms.
2 Bilaterally symmetrical.
3 Triploblastic, with well-developed organ systems.
4 Excretory system generally of flame cells and ducts.
5 Nervous system includes a brain.
6 A coelom is lacking, the spaces between organs being filled with parenchyma.
7 Generally hermaphrodite. The ovary is frequently divided into a germarium producing ova, and a vitellarium producing cells which contain yolk and shell-forming substance.

Larval forms

TURBELLARIA	DIGENEA	ASPIDOGASTREA
Where present, a Müller's larva or Götte's larva	Miracidium Sporocyst Redia Cercaria Metacercaria	Cotylocidium

MONOGENEA	CESTODA
Oncomira-cidium	Coracidium Procercoid Plerocercoid Hexacanth (Oncosphere) Cysticercoid Cysticercus

Metamorphosis

Turbellaria. Generally a juvenile worm hatches from the egg. A ciliated free-swimming larval form is present in a few species.

Digenea. The miracidium (Fig. 7.1A) which hatches from the egg has locomotory cilia, eyespots and an anterior penetration organ for entry into the molluscan intermediate host. After entry the larva loses its cilia and in the digestive gland of the host develops into a hollow sac or sporocyst containing germinal cell balls. These cell balls give rise to another larval stage, the redia, with pharynx and intestine, appendages and frequently a birth pore for escape of larval forms. The redia also contains germinal cell balls which develop into tailed cercariae with a digestive tract and suckers (Fig. 7.1B). The cycle from miracidium to cercaria varies in different species, with the repetition or omission of a stage. The multiplication of larval forms within the mollusc host is termed polyembryony.

The free-swimming cercaria escapes from the mollusc host. It may enter the definitive host or a second intermediate host (where it encysts to form a metacercaria). When the second intermediate host is eaten by the definitive host the larva is released and develops into the adult form.

Aspidogastrea. After invasion of the host, the ciliated or non-ciliated larva undergoes slow development (no metamorphosis) to the adult, with elaboration of the ventral sucker and the development of genitalia.

Monogenea. The oncomiracidium has cilia, eyes, a gut and a posterior hooked haptor. There is rarely an intermediate host in the life-cycle and the larva directly infects the definitive host. In metamorphosis the cilia are lost, there are changes in the form of the haptor, and genitalia are acquired.

Cestoda. The newly-hatched larva is generally 6-hooked. It may be a coracidium (Fig. 7.2A) (free-swimming, or not hatching from the egg till in the gut of the intermediate host) or a membranated hexacanth embryo or oncosphere (Fig. 7.2C). The

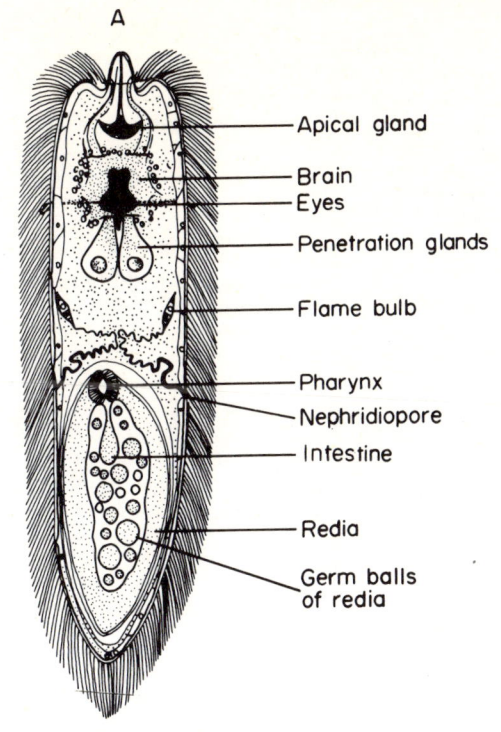

Fig. 7.1 Digenean larvae. A. Miracidium of *Parorchis*, containing a fully-developed redia (after Rees). B. Cercaria of *Alaria* (after Bosma).

hexacanth is released only after digestion of the membranous envelope in the gut of the intermediate host.

The cestode larva enters the body cavity of the intermediate host, where its anterior end grows. The coracidium develops into a *procercoid* larva (Fig. 7.2B) which develops further only when the intermediate host is eaten by the definitive host. If there is a second intermediate host the procercoid, on penetrating its body cavity, develops into an unsegmented *plerocercoid* larva with a scolex, which only reaches maturity when this new host is eaten by the definitive host.

In the body cavity of the intermediate host, the hexacanth develops a scolex and becomes a *cysticercoid* larva (Fig. 7.2D) if the scolex is retracted into the larval body, or a *cysticercus* (Fig. 7.2E) if the scolex is introverted. In the definitive host protrusion or eversion of the scolex occurs, the larva attaches, and strobilization begins.

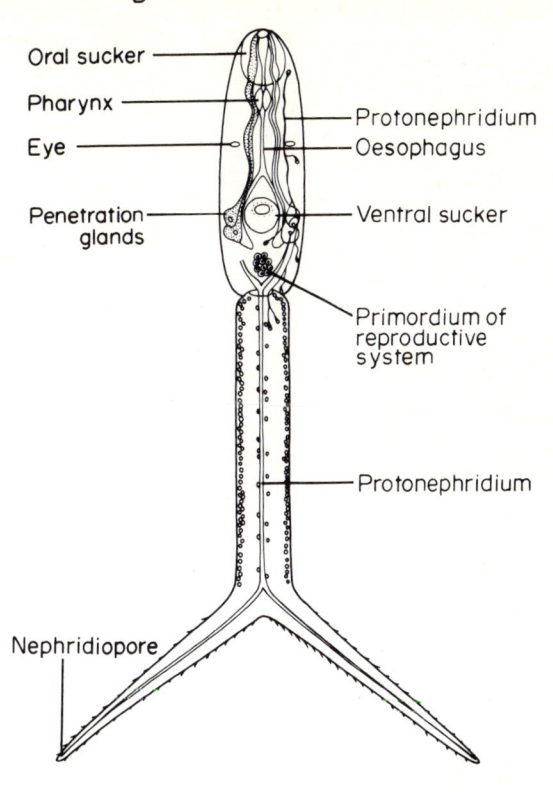

Adult body form

The Platyhelminthes are metazoan worms, generally dorso-ventrally flattened, and bilaterally symmetrical. There is an anterior end (but no well-defined head), which often bears sense organs. The mouth may be anterior or ventral. Attachment organs are best developed in parasitic forms.

A tissue called parenchyma (of mesodermal origin) fills the spaces between organs, between epidermis and gastrodermis.

There is no hard general body skeleton, but the parenchyma has a hydrostatic skeletal function.

1 *Class* TURBELLARIA e.g. *Dendrocoelum* (Fig. 7.3), *Mesostoma, Temnocephala, Polycelis*. Mostly free-living; with ciliated epidermis whose cells contain rhabdites; gut present except in Acoela; flame cells (protonephridia) absent in some species; in the remainder excretory pores vary in number

Phylum PLATYHELMINTHES

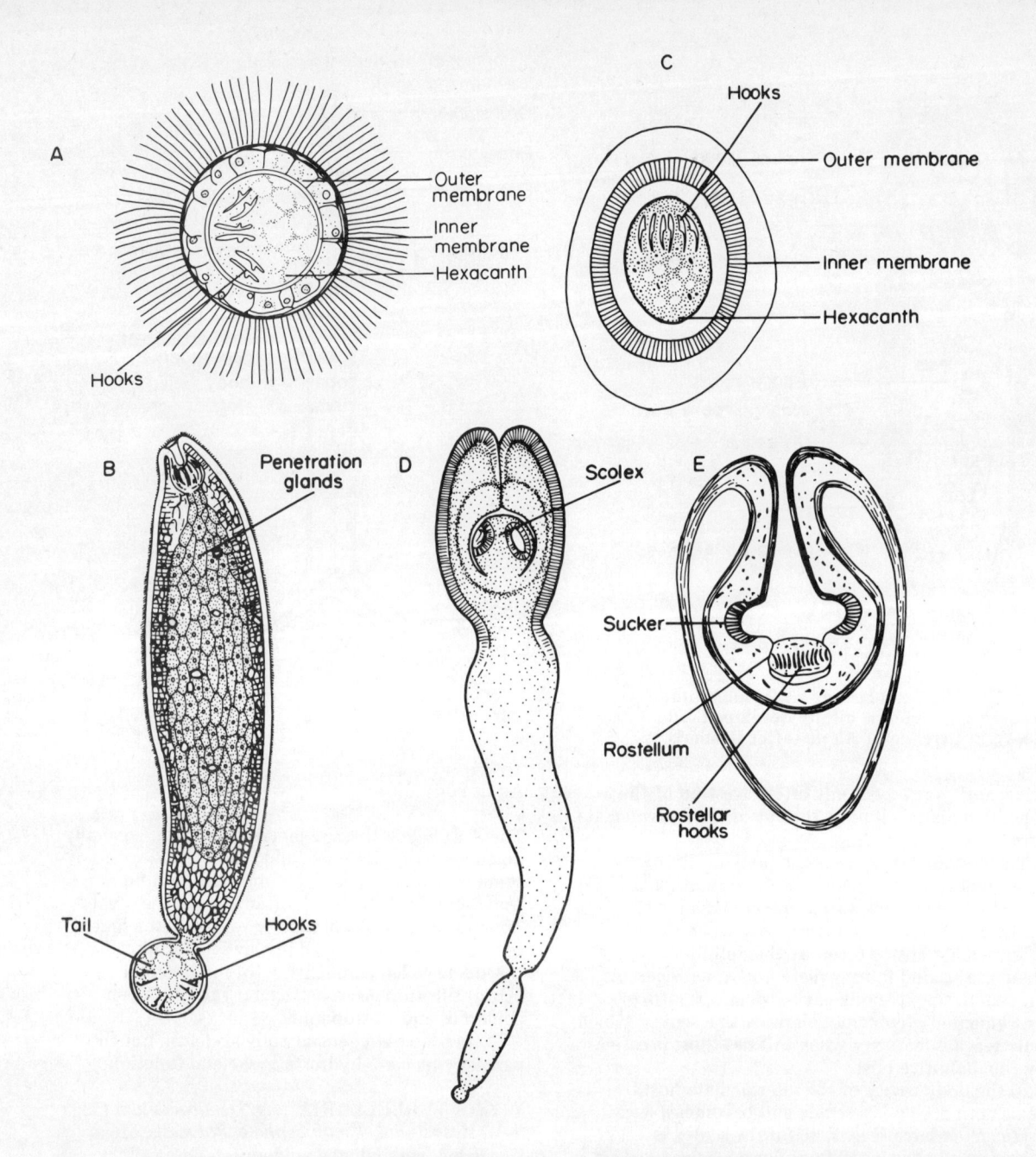

Fig. 7.2 Cestodan larvae. A. Ciliated coracidium of *Diphyllobothrium latum*. B. Procercoid of *Diphyllobothrium latum* (A and B after Rosen). C. Hexacanth of *Hymenolepis gracilis* (after Joyeux and Baer). D. Cysticercoid of *Hymenolepis diminuta* (after Joyeux). E. Cysticercus with introverted scolex (after Hyman).

Phylum PLATYHELMINTHES

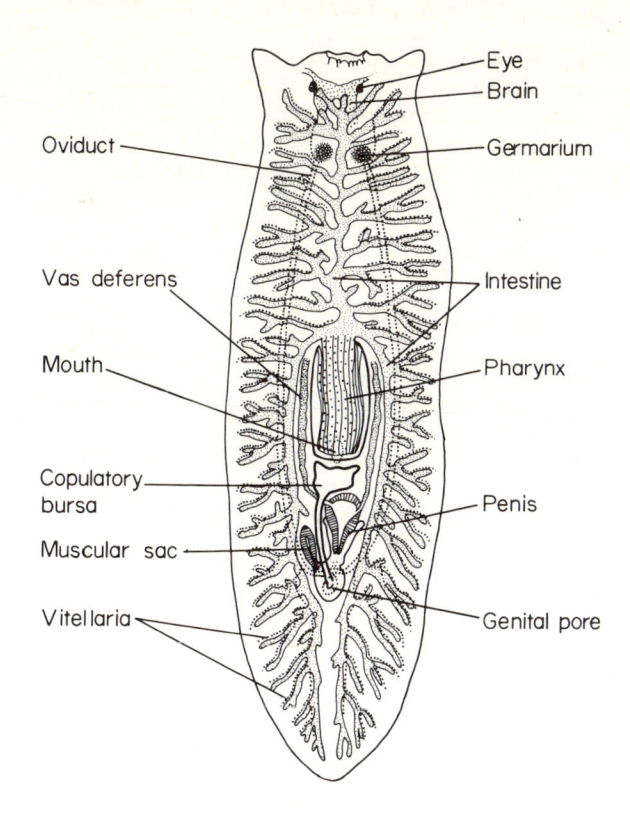

Fig. 7.3 Semi-diagrammatic dorsal view of the turbellarian *Dendrocoelum lacteum*, testes omitted (after Beauchamp).

and position; some groups show spiral cleavage, the remainder irregular cleavage.

2 *Class* DIGENEA e.g. *Fasciola, Schistosoma, Opisthorchis.* Endoparasitic; non-ciliated epidermis lacks rhabdites; gut always present, typically with an oral sucker (Fig. 7.4); adult has a single posterior excretory pore; generally a simple, unarmed ventral sucker for attachment to host; two to four-host life-cycle.

3 *Class* ASPIDOGASTREA e.g. *Aspidogaster.* Endoparasitic; non-ciliated epidermis lacking rhabdites; oral sucker lacking and intestine simple; adult has a single posterior excretory pore; for attachment there may be a number of separate ventral suckers, but generally there is a single large sucker sub-divided into numerous alveoli (Fig. 7.5); generally a single-host life-cycle.

4 *Class* MONOGENEA e.g. *Diclidophora, Polystoma, Gyrodactylus.* Generally ectoparasitic; epidermis without cilia or rhabdites; gut generally present, with a simple mouth, a mouth with an oral sucker, or a buccal cavity with a pair of eversible suckers; the pair of excretory pores are anterior and lateral; there is a posterior, terminal, generally armed haptor for attachment (Fig. 7.6A); single-host life-cycle.

5 *Class* CESTODA e.g. *Diphyllobothrium, Taenia, Echinococcus.* Endoparasitic; body generally composed of a number of proglottids or segments (Fig. 7.7); epidermis with microvilli but not cilia or rhabdites; gut absent; adult at first has a single posterior excretory pore, with the loss of

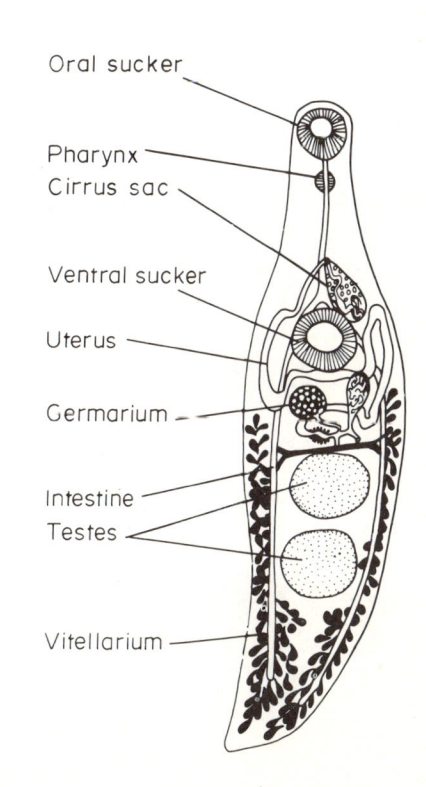

Fig. 7.4 A digenean, *Allocreadium isosporum* (after Looss).

Phylum PLATYHELMINTHES

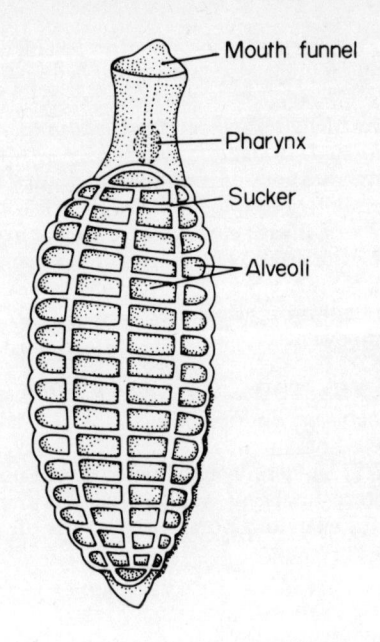

Fig. 7.5 *Aspidogaster conchicola* (after Monticelli).

proglottids number may increase; anterior attachment organ or scolex (Fig. 7.8) joined by a neck to the strobila of proglottids; generally hermaphrodite, each proglottid containing male and female reproductive organs; life-cycle with or without intermediate hosts.

Feeding

a   Free-living turbellarians are carnivorous. Parasitic forms with a gut feed on tissues and tissue fluids of the host. The Cestoda, which lack a gut, absorb nutrients from the intestine of the host.

b   Where a gut is present, an anterior or ventral mouth opens into a pharynx (a buccal cavity or a pre-pharynx may be present) which leads into a blind intestine. The intestine may be simple or divided into two, three or many limbs which may be subdivided. The pharynx may be muscular and protrusible or withdrawn into the body. Where withdrawn there is an oral sucker, or a pair of buccal suckers in the pre-pharyngeal cavity.

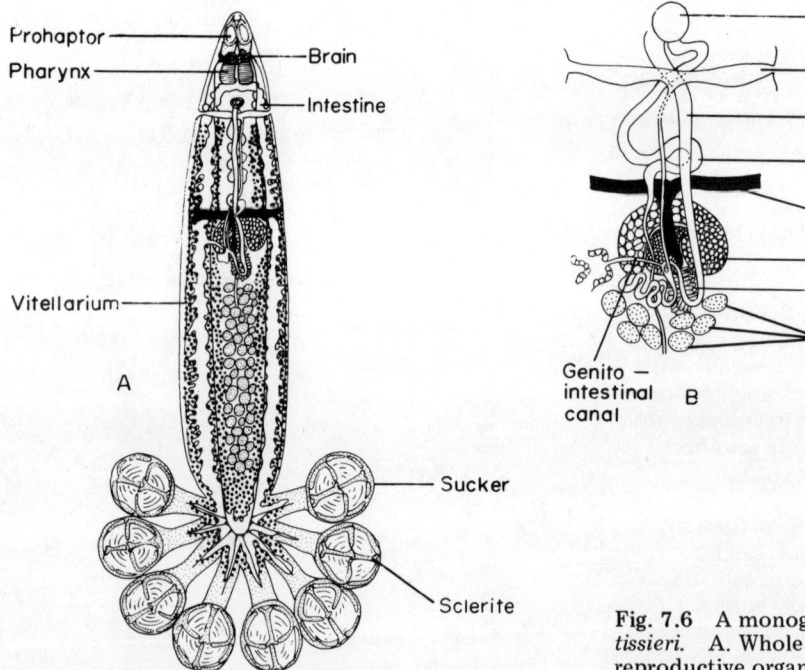

Fig. 7.6 A monogenean, *Diclidophoropsis tissieri*. A. Whole animal. B. Diagram of the reproductive organs (A and B after Gallien).

Phylum PLATYHELMINTHES

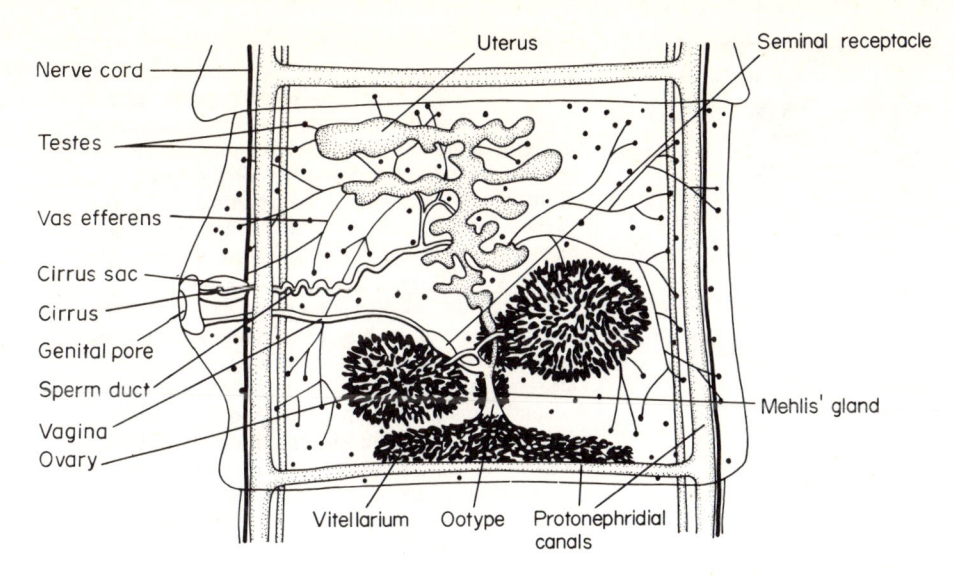

Fig. 7.7 A proglottid of the cestode *Taenia* (after Noble and Noble).

A protrusible pharynx may be used to perforate the skin of large prey and suck out tissues (e.g. some turbellarians). Where oral suckers are present they are generally used to suck in tissue fluids and blood.

Digestion is both extracellular and intracellular, sometimes beginning outside the body of the worm (e.g. in strigeid digeneans). Where a gut is present it is the site of enzyme production and release, digestion and absorption. In the Cestoda nutrients are absorbed from the host intestine (in some species at the attachment organ as well as from the lumen) by diffusion or some active process.

Osmoregulation/excretion

a The system of flame cells (protonephridia) and ducts characteristic of platyhelminths is not present in acoelan and polyclad turbellarians.

Each flame cell consists of a nucleated cell body produced in one region into a cylinder whose extracellular lumen encloses a large group of cilia, united so that they beat together (Fig. 7.9). The lumen is continuous with a duct which unites with similar ones to form a system that converges on the excretory pores. The ducts may contain ordinary cilia.

b The system of flame cells and tubules is presumed to be involved in osmoregulation.

The physiology of excretion has been little studied and the role of the flame cells and tubules is not known. The main end products of nitrogen metabolism are ammonia, urea and uric acid.

Movement

a Types of locomotion are swimming, creeping, movement by muscular waves and leech-like movement.

b Swimming is generally accomplished by means of cilia (e.g. in larvae and most acoelan turbellarians). In cercariae the tail is used.

In creeping (e.g. larger turbellarians) cilia play an important role though muscles may also be involved.

Polyclad turbellarians move partly or entirely by muscular waves.

Leech-like progression is achieved by means of muscles and anterior and posterior attachments (suckers and hooks).

c The integumentary muscles are sub-epidermal and typically consist of an outer layer of circular

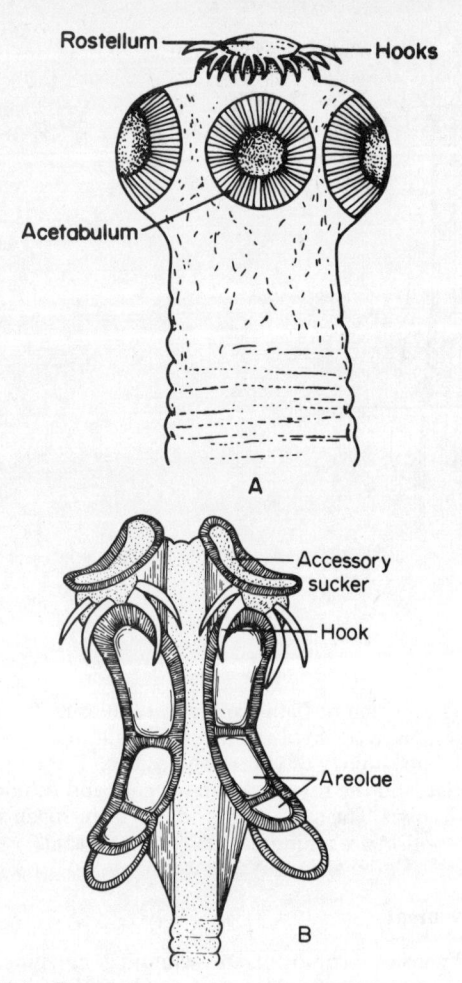

Fig. 7.8 Examples of cestodan scolices. A. Scolex of *Taenia solium*. B. Scolex of *Calliobothrium* (A and B after Southwell).

muscles and an inner layer of longitudinal ones. In addition, however, there may be another layer between the circular and longitudinal ones (e.g. some turbellarians and monogeneans), or on the inside of the longitudinal layer (e.g. some digeneans). Cestodes have, in addition to longitudinal and circular muscle layers, a secondary mesenchymal musculature in the parenchyma, of longitudinal, transverse and dorso-ventral fibres. Dorso-ventral strands are also present in many acoelans.

Muscle physiology has been little studied.

Co-ordination

a i A central nervous system is present, with an anterior bilobed brain giving rise to a number of anterior branches and a variable number of paired posterior cords connected by commissures. Construction varies with group and species, the highest level of development occurring in free-living forms. The cellular differentiation of the CNS is greatest in the Turbellaria.

The peripheral nervous system varies in plan with the group and may consist of deep and superficial plexuses connecting with the CNS. The pharynx, attachment organs and genitalia have nerve plexuses.

Sense organs are best developed in free-living forms. In the Turbellaria they include eyes, statocysts, specialized patches of sensory epithelium, scattered photoreceptors, tactile, chemo- and rheoceptors. The latter are sensitive to water currents.

ii The physiology of the nervous system has been little studied. In Turbellaria the brain is apparently important in locomotion, food-seeking and recognition but in its absence the longitudinal cords can mediate some locomotive, righting and avoidance reactions.

Except for the pharynx, there is no general nerve net to mediate responses peripherally though in polyclads the peripheral plexus may mediate some simple reflexes locally. The nerve plexus of the

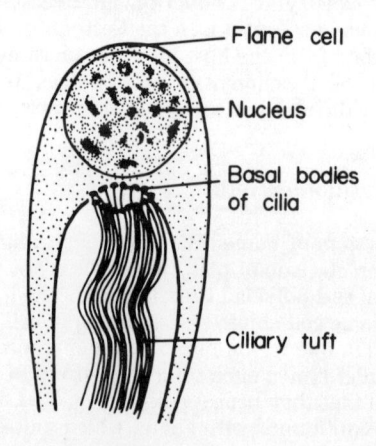

Fig. 7.9 A flame cell of a planarian (Turbellaria) (after Willey and Kirschner).

40   Phylum PLATYHELMINTHES

pharynx is able to co-ordinate feeding movements, when isolated from the body. Except during feeding this action is inhibited normally by a connection with the ventral cord.

b There is some evidence of wound hormones in Turbellaria and substances active in regeneration. Cestodes may regulate proglottid shedding by specific chemicals. Neurosecretory cells have been reported in some groups.

Respiration

a There are no special organs.

b Respiratory exchange occurs across the general body surface. Free-living and ectoparasitic forms show aerobic respiration. Adult digeneans are generally considered to be facultative anaerobes. In the Cestoda anaerobic metabolism predominates but species so far examined experimentally, have been found to utilize oxygen when available.

Circulation/coelom

a There is no circulatory system, haemocoel or coelom.

b The parenchyma and its fluid-filled spaces, play a role in internal transport.

c Haemoglobin has been reported in some digenean species but its role is not known.
   In certain digenean families a so-called 'lymphatic' system (mesenchymal) exists but its role (circulatory or otherwise) is unknown.

Reproduction

a Most platyhelminths are hermaphrodite. The following account of the reproductive organs is general and variations may be found in different groups. Fig. 7.10A is a general plan of helminth reproductive organs.
   The male reproductive organs are testes, their collecting ducts (the vasa efferentia and deferentia), and an intromittent organ for sperm transfer (the cirrus, if eversible, or penis, if protrusible and muscular). Part of the vas deferens is often dilated to form a seminal vesicle for sperm storage. Accessory structures called prostate glands may be associated with the copulatory apparatus (Fig. 7.10B, C)
   The female gonad may be an ovary producing yolky eggs, or may be divided into a germarium producing ova and a vitellarium producing cells containing yolk and shell-forming substance. From the vitellarium and ovary the vitelline duct and oviduct unite to form an ovo-vitelline duct, which has a dilated region, the ootype, where zygotes and vitelline cells are combined to form shelled eggs. The ootype is surrounded by Mehlis' gland. The oviduct may have a dilatation for sperm storage, the receptaculum seminis, which may communicate with the exterior by a vagina (in Digenea Laurer's canal is present (Fig. 7.10D)). The oviduct or ovo-vitelline duct leads into a uterus which generally opens, with the male penis or cirrus, into a common genital atrium. In some cestodes the uterus is blind. In many species (especially where a functional vagina is absent) the distal end of the uterus is modified as a metraterm (Fig. 7.10B) for reception of the male copulatory organ.

b *Sexual reproduction.* Cross or self-fertilization may occur. The cirrus or penis is inserted into the metraterm or vagina or Laurer's canal. Spermatozoa are stored in the receptaculum seminis until required for fertilization of ova. Fertilized ova and vitelline cells are assembled into eggs in the ootype, and then the shelled eggs pass into the uterus, from which they are released. In tapeworms, where the uterus has no external opening, ripe proglottids with gravid uteri are detached from the strobila and passed out with the host faeces.
   In some Turbellaria cleavage of the zygote is spiral, in other flatworms it is said to be irregular.

c *Asexual reproduction* may occur by:

i fragmentation and regeneration (e.g. some turbellarians);

ii transverse fission (e.g. some turbellarians);

iii polyembryony (e.g. in digenean larvae);

iv budding (e.g. in cysticercoids and cysticerci of cestodes).

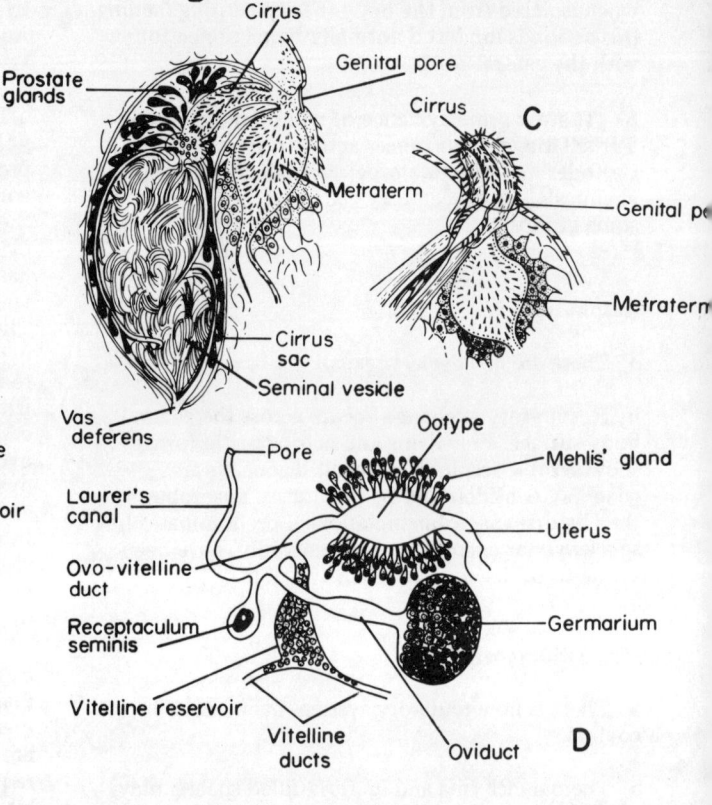

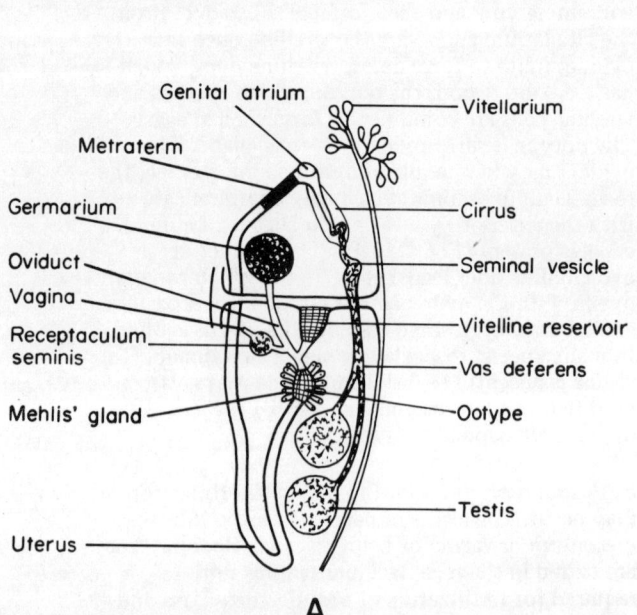

Fig. 7.10 Reproductive organs of platyhelminths. A. General plan (after Llewellyn). B. Male copulatory apparatus of a digenean with cirrus introverted. C. The same, with cirrus everted (B and C after Looss). D. Plan of the female reproductive system of a digenean (after Hyman).

Regeneration

In turbellarians which reproduce asexually (see above) regenerative powers are high. Regeneration is correlated with polarity so that, for example, the anterior portion of an excised piece of an animal will regenerate an anterior end. Cells of the parenchyma are involved in the production of new tissue.

Parasitism

The phylum Platyhelminthes contains important groups of animal parasites, with life-cycles ranging from simple, with a single host, to complex with one or more intermediate hosts and a final definitive host, which is nearly always a vertebrate.

The degree of host specificity (restriction of the parasite to a particular host) is variable. The site which the mature parasite occupies in or on the host is specific.

Morphological attributes which are important in the parasitic mode of life are:

i penetration organs;

ii attachment organs;

iii proliferation of the gonads to increase reproductive capacity.

Phylum PLATYHELMINTHES

Free-living larval stages with the ability to seek out and infect new hosts are also important to parasitic species.

Organs or organ systems which are less essential to parasites than to free-living forms tend to be reduced, e.g. sense organs, locomotor organs, digestive tract.

A knowledge of life-cycles is of particular importance in those species which are facultative or obligatory parasites of man and domestic animals.

References

Grassé P.P. (Editor) 1961. *Traité de Zoologie* Vol. IV (1). Masson et Cie, Paris.
Smyth J.D. 1966. *The Physiology of Trematodes*. Oliver and Boyd, Edinburgh and London.
Smyth J.D. 1969. *The Physiology of Cestodes*. Oliver and Boyd, Edinburgh and London.

8 Phylum NEMERTINI

600 species described. Size ranges from minute to very long (several metres).

Characteristics

1 Most marine, some terrestrial.
2 Vermiform, elongate, flattened and unsegmented.
3 Have a characteristic and unique organ, the proboscis, which lies dorsal to the gut.
4 Possess a ciliated epithelium.
5 Alimentary canal has a mouth and anus.
6 A blood vascular system is present.
7 Acoelomate, having no body cavity. The body is instead packed with parenchymatous tissue.
8 The sexes are separate with simple gonads repeated serially along the length of the body.
9 Larval form, where present, is planktonic — the pilidium.

Larval form

Most nemertines develop straight from the egg. Some give rise to a pelagic larva, the pilidium (Fig. 8.1). The apical tuft is composed of cilia which are usually immobile, and probably represent a sensory area, though whether it be mechanically or chemically sensitive remains to be demonstrated. The lappets of the larva are heavily ciliated.

In planktonic larvae the juvenile form of the nemertine develops within a specialized area of the pilidium known as the amnion. When ready for hatching the remains of the larva are resorbed or discarded and the immature worm is fully formed.

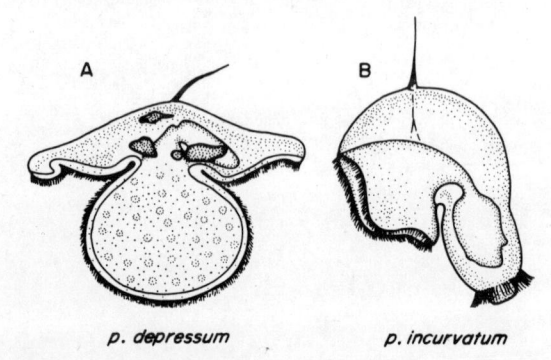

Fig. 8.1 Two forms of pilidium larva (names do not apply to species) (after Jägersten).

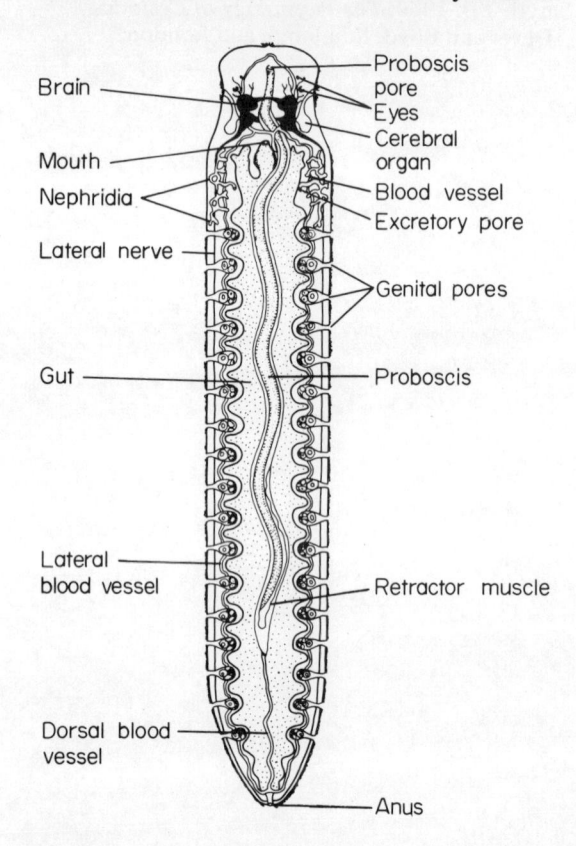

Fig. 8.2 Basic anatomy of a nemertine (after Kükenthal).

44   Phylum NEMERTINI

Lineus produces a creeping larva known as the larva of Desor.

Metamorphosis

Within the pilidium five ectodermal sacs form that eventually link up to create a cavity around the larval gut. The imaginal discs forming the floor of the sac (amnion) join together to form the secondary ectoderm and the juvenile grows from them and becomes fully formed within the amnion.

Adult body form

The adult nemertine is an elongate animal, covered with motile cilia and possessing an extremely long dorsal proboscis (Fig. 8.2 gives the basic details; Fig. 8.3 indicates major differences in the four orders).

There are two Classes.

1 ANOPLA, in which the mouth lies just posterior to the position of the brain, the nerve cord lies just below the epidermis (Fig. 8.4) and the proboscis bears no stylets.

Order Palaeonemertini has 2 or 3 layers of body musculature of which the innermost is circular, e.g. *Carinoma, Tubulanus*.

Order Heteronemertini has 3 layers of body musculature, the innermost being longitudinal, e.g. *Lineus, Cerebratulus*.

2 ENOPLA, in which the mouth lies anterior to the brain, the nerve cord is deep within the body wall muscles (Fig. 8.4) and the proboscis may be armed (with stylets).

Order Hoplonemertini: proboscis armed, e.g. *Amphiporus*.

Order Bdellonemertini: proboscis unarmed, with a posterior sucker and commensal in bivalves, e.g. *Malacobdella*.

Body form is maintained internally by the parenchymatous tissue.

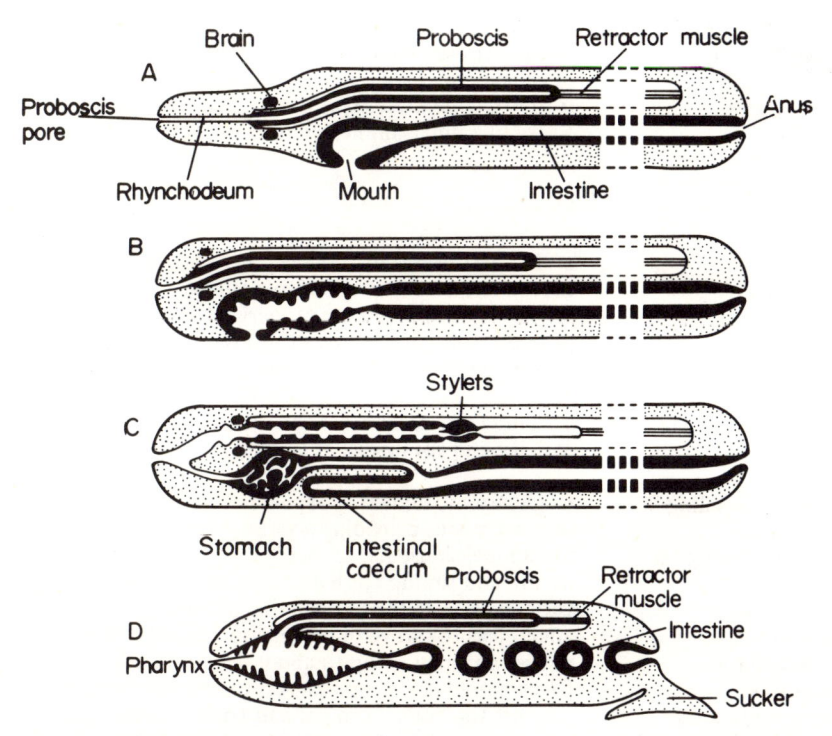

Fig. 8.3 Relationships of gut and proboscis in the orders of nemertines. A. Palaeonemertini. B. Heteronemertini. C. Hoplonemertini. D. Bdellonemertini (A—D after Gibson).

Phylum NEMERTINI

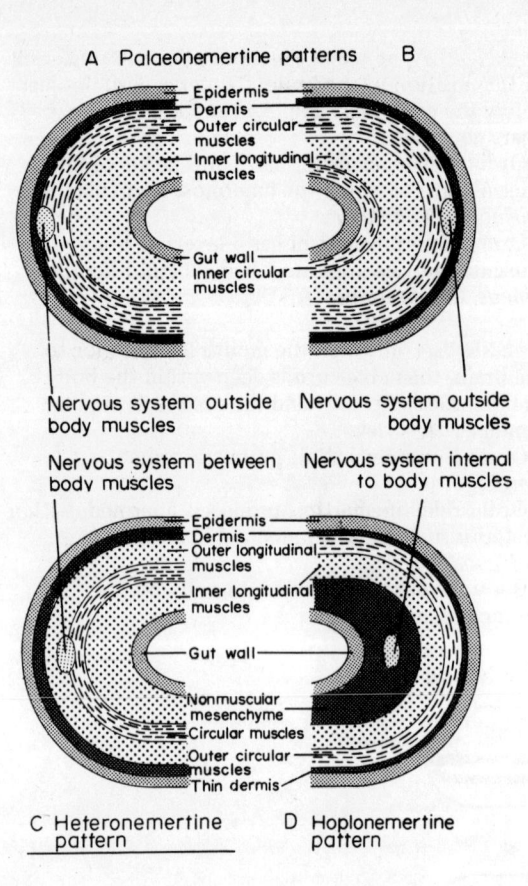

Fig. 8.4 Relationships of nerve cords to body wall musculature. A & B. Palaeonemertini. C. Heteronemertini. D. Hoplonemertini (A–D after Russell-Hunter).

Feeding

a Nemertines are carnivorous predators. Feeding in nearly all cases is by means of the proboscis.
Malacobdella is perhaps an exception, feeding on plankton filtered by its host.

b The proboscis lies in a cavity dorsally (the rhynchocoel) (Fig. 8.5) and when retracted is turned inside out upon itself; extrusion is accomplished by contraction of body wall muscles exerting pressure on rhynchocoel walls and consequently raising the hydrostatic pressure within the rhynchocoelic cavity. Upon contact with the prey a glandular region secretes mucus and possibly a poison which can be introduced into the prey via punctures made by the stylets (when present). In forms without stylets the proboscis may be used as a lassoo, coiling around the food object. Return of the proboscis to the rhynchocoel is by a retractor muscle or by lowering of internal pressures within the rhynchocoelic villus.

The alimentary canal is straight though some species have pouches arranged along the length of the tube; mouth (anteriorly) and anus (posteriorly) are virtually terminal in position.

Osmoregulation/excretion

a There is an excretory system composed of numerous flame cells. These are connected to longitudinally-arranged lateral excretory canals. The canals open by one or several pores along the body. The blood system in larger nemertines may be closely associated with the flame cells.

b The activity of the flame cells and their anatomy seems correlated with the osmoregulatory abilities of the animal, being best developed in terrestrial forms (*Geonemertes* sp.). Flame cell cilia are most active in animals immersed in fresh water.

Movement

a Locomotion is of three types; one is slow gliding movement over the substratum (up to 1 cm/min); a second is fast movement by peristaltic contraction of body wall muscle and a third is swimming which occurs in some of the more active forms.

b These movements are the result of (1) ciliary activity and (2) muscular activity.

c Ciliary movement is dependent upon a copious flow of mucus providing a watery medium around the animal. Muscular movement is peristaltic in nature and similar to that of earthworms, with various parts of the body acting as temporary anchors as the wave of contraction passes. Speed of movement may be up to about 10 cm/min. In *Rhynchodemus* myopodia (false feet) are formed that lift the body from the substratum. Muscular movement is limited by the structure of the geodesic fibres of the body wall coverings.

46 Phylum NEMERTINI

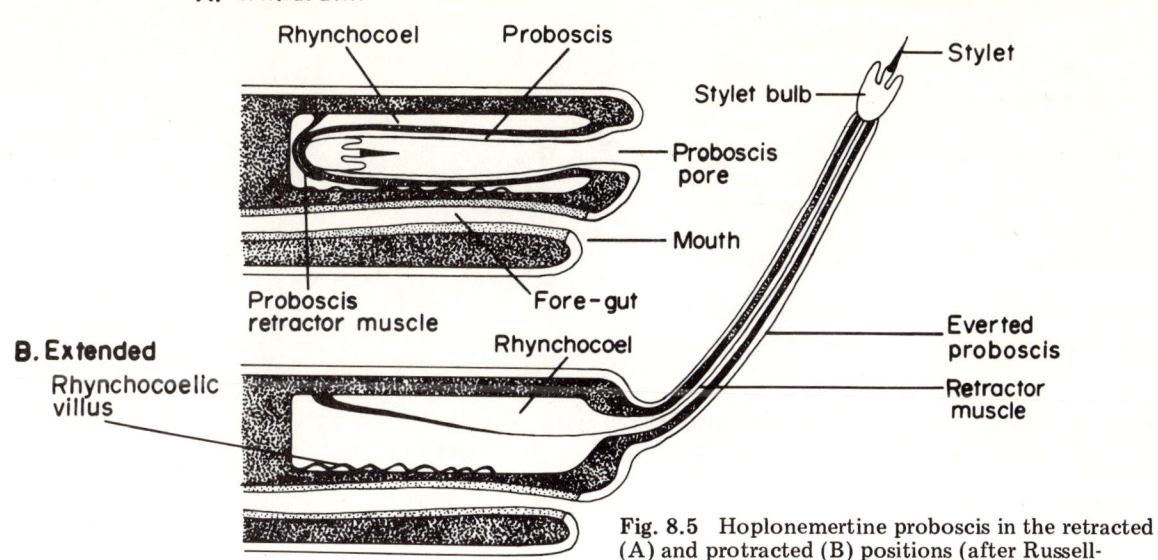

Fig. 8.5 Hoplonemertine proboscis in the retracted (A) and protracted (B) positions (after Russell-Hunter).

Co-ordination

a i Nervous system: there is a nerve net that becomes complex in advanced forms. A pair of cerebral ganglia are formed above the mouth and from these, lateral cords run along the length of the body. The relative disposition of the lateral cords and the muscle layers is a diagnostic feature used in classifying the nemertines (Fig. 8.4). Some forms, e.g. *Cerebratulus*, have giant fibres which may be associated with a swimming habit.

Sense organs include eyes (variable in number), mechanoreceptors, and a putative chemoreceptor, the cephalic organ. This is at the base of a slit-like fold on the head of the nemertine which brings the external environment in close contact with the cerebral ganglia. It is well-developed in some active nemertines but is not found in *Neuronemertes* (a pelagic abyssal form) and is degenerate in *Malacobdella*; this may be a reflection of the comparative stability of the environments of these genera.

ii Circular muscle contracts under the action of acetylcholine, longitudinal muscle to adrenaline, and retractor muscle to oxytocin.

b No hormones have been isolated from nemertines. This is the lowest animal group with a developed blood system however, which perhaps suggests that endocrine systems may be present. Neurosecretion has been reported. A secretory function has been described for the cephalic organ.

Respiration

a There are no special adaptations.

b Diffusion occurs across the epidermis. *Cerebratulus* demonstrates rhythmic intake of water to the oesophagus.

Circulation/coelom

a Nemertines are acoelomate, but possess a blood system of closed vessels. These lie within the parenchyma, and consist of a few (1 to 3) longitudinal vessels connected by transverse vessels.

b The blood is usually colourless and is propelled by body wall movements, although some of the vessels are muscular and contractile.

Phylum NEMERTINI

c Haemoglobin is present in some species. Its function is unknown.

Reproduction

a The sexes are generally separate, but some species are hermaphrodite. The gonads are paired sacs situated along the length of the body, lying between the pouches of the alimentary tract. They develop a short duct to the exterior when ripe.

b *Sexual reproduction.* External fertilization occurs, cleavage is spiral and determinate.

c *Asexual reproduction.* Some nemertines reproduce by fragmentation and regeneration of the pieces. A few common intertidal nemertines may reproduce by fragmenting in spring and summer, and sexually when water temperatures begin to fall.

Reference

Gibson R. 1972. *Nemerteans.* Hutchinson University Library, London.

9 Phylum GASTROTRICHA

Microscopic animals, the group is composed of about 150 species.

Characteristics

1 Aquatic animals, both fresh-water and marine species.
2 Lacking segmentation, worm-like.
3 Possess characteristic ciliated bands.
4 Have superficial bristles, scales and spines.
5 With pair(s) of adhesive tubes.
6 Protonephridia may be present.
7 No coelom but some signs of pseudocoel formation.

Larval form

The hatched egg releases a juvenile gastrotrich that takes about three days to become mature.

Metamorphosis

There is no larval stage as such, and hence no metamorphosis.

Adult body form e.g. *Chaetonotus, Cephalodasys*.

Gastrotrichs are elongate, ventrally flattened and spinous. There is a head, neck and trunk, whose relative proportions differ from species to species (Fig. 9.1). They bear ciliated tracts, especially on the head and ventral side of the trunk, although in some species these are absent. The body is covered with a thin cuticle from which scales, bumps and spines are formed. The typical adhesive tubes are cuticular cylinders that are adjustable by muscles and produce glandular secretions. Internally the epidermis is syncytial, and the muscles are not arranged in definite layers but singly or grouped in a variety of orientations. The mouth is terminal (Fig. 9.2).

Feeding

a Microphagic, suctorial or ciliary feeders.

b There is a muscular unarmed pharynx which is lined by cuticle. The gut is tubular and straight, opening at a sphinctered anus.

In most forms feeding is accomplished by the sucking action of the pharynx; in some, ciliary currents bring particles to the mouth.

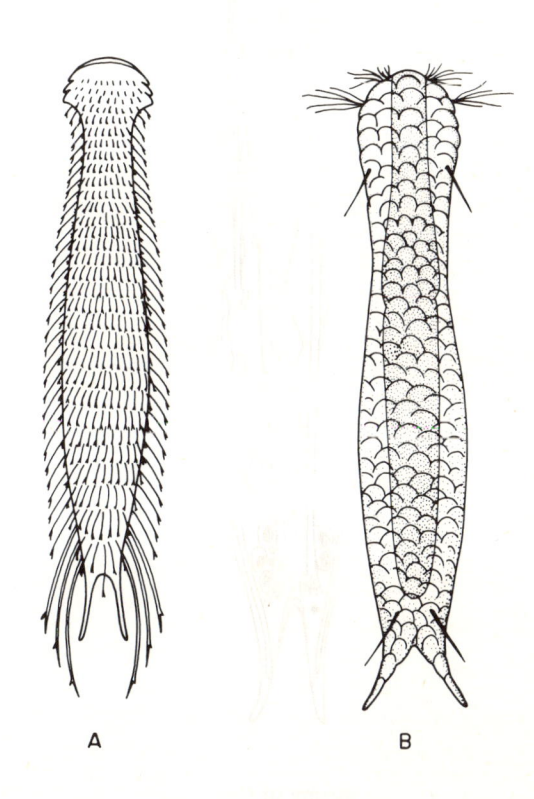

Fig. 9.1 A. *Chaetonotus zelinkae* (after Grunspan). B. *Lepidoderma squamatum* (after Zelinka).

Osmoregulation/excretion

a Protonephridia are found in one order (Chaetonotoidea). They are a pair, each consisting of one flame cell connecting to a long tubule that opens ventrally.

b The physiology is unknown.

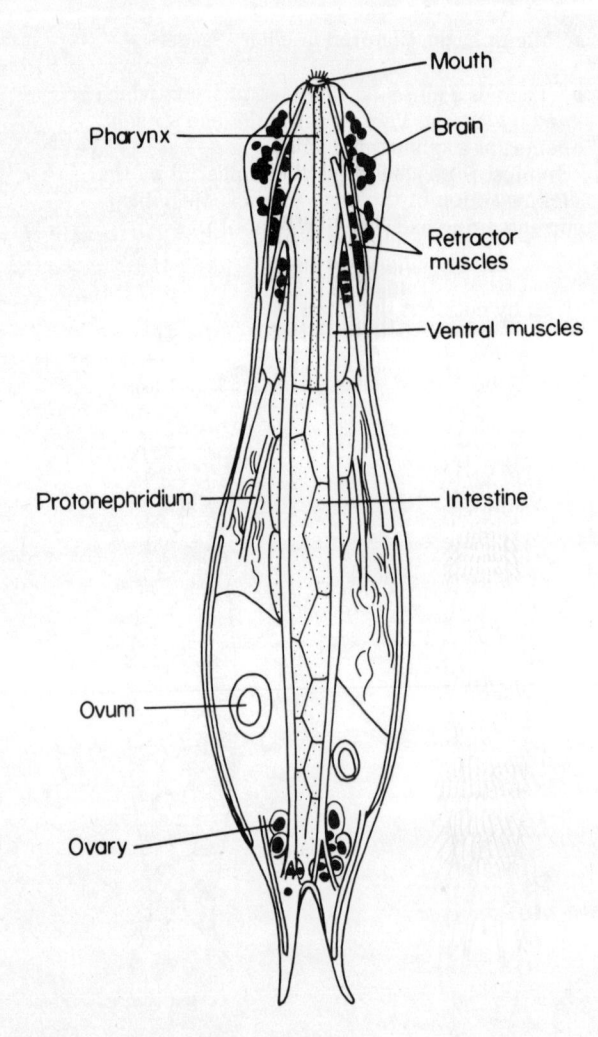

Fig. 9.2 Internal anatomy of *Chaetonotus* (after Zelinka & Remans).

50   Phylum GASTROTRICHA

Movement

a Swimming or gliding and occasionally looping movements (with temporary adhesion) are seen.

b Cilia provide the main motive power, but muscle systems provide for some changes in body shape.

c The physiology is unknown.

Co-ordination

a *i* The nervous system is more pronounced than that of kinorhynchs. There is a large brain and a pair of lateral nerves that pass along the body. Sense organs are stiff cilia and hairs, pigmented ocelli, presumptive chemoreceptors, and possibly equilibrium receptors.

ii The physiology is unknown.

b Hormones are not known.

Respiration

a There are no respiratory organs.

b Respiratory exchange must be a general bodily function.

Circulation/coelom

a The group is pseudocoelomate. The pseudocoel is very small.

b Any fluid movement must be small and due to muscle action in the gut and body wall.

c There are few amoebocytes.

Reproduction

a In Chaetonotoidea only females are known, one or two ovaries being present. Among other groups hermaphroditism is common, but some species may have separate sexes. The ovary may possess a seminal receptacle for sperm storage. The male system is one or a pair of testes, each with a sperm duct opening by a gonopore, which may be close to or confluent with the female gonopore.

b   A penis-like organ is found in one or two species which suggests copulation may be practised. In groups with no males, parthenogenesis is the only method of reproduction. The embryology is not well known.

Reference

Hyman L. H. 1951. *The Invertebrates*, Vol. III. McGraw-Hill, New York.

10 Phylum KINORHYNCHA

Microscopic animals, about 100 species in all.

Characteristics

1. Exclusively marine.
2. A number of larval types have been described.
3. There are no cilia.
4. The body is superficially segmented with 13 or 14 segments or zonites.
5. Possess a retractile head bearing spines.
6. There is one pair of protonephridial tubules, each having one flame cell.

Larval form

Early larvae show little resemblance to adults, but there are differences amongst the larvae which seem to depend upon the stage at which hatching occurs (the numbers of segments or zonites vary). Juvenile stages also occur in some species.

Metamorphosis

Larvae metamorphose by a series of moults, the adult form gradually developing as more zonites are added. Juveniles differ from adults only in having softer cuticle, few segments, and fewer teeth, spines and bristles.

Adult body form e.g. *Echinoderes, Echinoderella*.

The body is formed by a head, a neck (1 zonite), and a jointed trunk with terminal spines. There are no surface cilia. The total number of segments is 13 or 14; the head accounts for one, the neck region for one, and the trunk for 11 (save *Campylodeses* which has 12). The head is retractable into the neck region, bears spines and carries a terminal mouth. The trunk zonites are all similar, covered by hardened cuticular plates joined by flexible regions at segmental boundaries. There are many spines and bristles, and a mid-ventral groove. A pair of adhesive organs, with gland cells, are placed anteriorly on the trunk.

Internally the epidermis is syncytial. Muscles are arranged segmentally, and some are organized to retract the head, whilst others anteriorly bring about protrusion of the head. A pseudocoel exists and is fluid-filled, acting as a hydrostatic skeleton. There are no segmental septa.

Feeding

a Kinorhynchs are suctorial grazers feeding on diatoms, detritus and fine particles.

b The alimentary canal opens at a terminal mouth; the buccal and pharyngeal regions are lined with thick cuticle; the main portion is a straight tube leading to a terminal posterior anus (Fig. 10.1A). Kinorhynchs ingest food by thrusting the eversible mouth cone out and sucking material in by means of the pharyngeal musculature.

Osmoregulation/excretion

a There is one flame bulb (multinucleate) in segment 10, opening via a sieve plate nephridiopore on segment 11.

b Physiology is not known.

Movement

a Movement is limited and cumbersome. The head is eversible and retractable and it may move from side to side. Progression is by use of the head and its spines as an anchor, drawing up the body, and then further extension of the head. No swimming occurs.

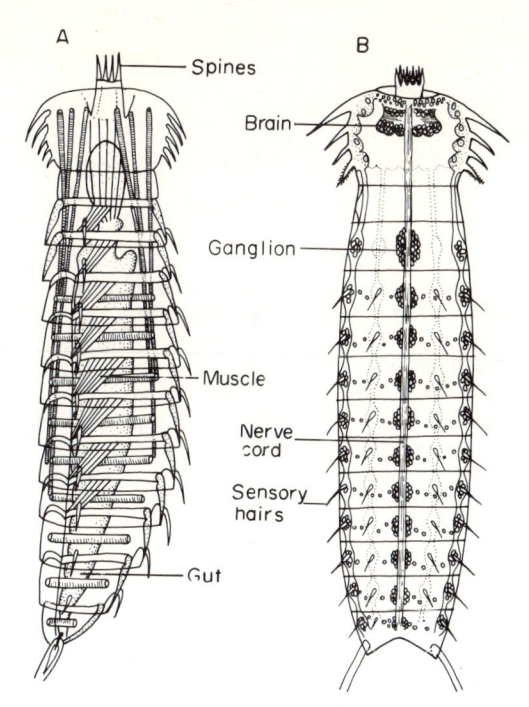

Fig. 10.1 Anatomy of a kinorhynch. A. Musculature and gut. B. Nervous system (A and B after Remans).

b No ciliary activity takes place, all movements being the function of the musculature acting on the fluid-filled central cavity.

Co-ordination

a *i* The nervous system is almost indistinguishable from the epidermis. A nerve ring circles the mouth cone and a ventral nerve cord from it runs the length of the body, with a ganglion in each zonite (Fig. 10.1B). Sense organs include eyes (in *Echinoderes*) and bristles along the body.

ii Physiology is not known.

b Hormones are unknown.

Respiration

a There are no special organs.

b Gaseous exchange is probably a general body surface phenomenon.

Circulation/coelom

a There is a fluid-filled pseudocoel in which amoebocytes are found.

b Any circulation of the fluid must occur as a consequence of body movements.

Reproduction

a The sexes are separate with the gonads existing as a single pair opening on the last segment.

b No details of fertilization, egg release and development are available.

Moulting

Moulting takes place, particularly during the larval stages. A new cuticle is formed beneath the old. Then the external layer, including the pharyngeal lining, is shed by a protracted struggle that splits the old covering and allows the escape of the animal.

Reference

Hyman L.H. 1951. *The Invertebrates*, Vol. III. McGraw-Hill, New York.

Phylum KINORHYNCHA

11 Phylum ROTIFERA

There are approximately 1500 known species, all very small.

Characteristics

1 Aquatic.
2 Males are generally smaller than the females.
3 The body is spherical or cylindrical, ending in a bifurcate foot.
4 The anterior part is modified to a ciliary organ, the corona or wheel organ.
5 The pharynx is armed with jaws (the mastax).
6 There is a well-formed cuticle.
7 Protonephridia are present.
8 Parthenogenesis is common.

Larval form

Female rotifers of free-swimming groups hatch as fully-formed adults, maturing in a few days. Sessile rotifers hatch as free-swimming juveniles which may be termed larvae.

Metamorphosis

In sessile forms attachment leads to degeneration of the eyes, loss of ciliature, elongation and development of the coronal lobes.

Adult body form e.g. *Epiphanes*, *Stephanoceros*, *Hydatina*.

Descriptions of rotifers apply to the females, the males being very much smaller and simpler. The integument is composed of two layers, the hypodermis and a chitinous secreted cuticle. This latter may be extremely stiff forming a lorica, and spinous or sculptured.

Rotifers have variable shapes, dependent on species and the nature of the lorica (Fig. 11.1A,B). Generally, however, there is a head bearing the wheel organ, a trunk and a foot (divided). The ventral side is usually flattened. No true segmentation is seen though in some species there are external signs of segmentation. The trunk bears palps or antennae. The body wall has no muscle layers but there are many single muscles in the body, arranged circularly and longitudinally. A cavity, the pseudocoel, exists internally; it is unlined by peritoneum and not crossed by mesenteries; it does not represent a true coelom, but rather the embryological blastocoel. The mouth opens into a pharynx that is armed with jaws (the mastax) that act in chewing (Fig. 11.1B).

The most intriguing feature of rotiferan organization is the constancy of cell numbers. The newly-hatched animal has the same number of cells as an older one; the position, shape and size of cells are fixed. Thus in *Epiphanes* 959 cells are found distributed as follows: coronal epidermis, 172; trunk and foot epidermis, 108; pedal glands, 19; circular muscles of the body wall, 22; retractor muscles, 40; brain, 183; peripheral nervous system, 63; mastax epithelium, 91; mastax musculature, 42; mastax nerve cells, 34; oesophagus, 15; stomach, 39; each gastric gland, 6; intestine, 14; vitellarium, 8; oviduct, 3; each protonephridium, 14. The total number of nuclei in those rotifers that have been studied is 900 to 1000.

Feeding

Three styles of feeding are exhibited by rotifers; the vortex, the grasping and the trapping types. In the first the corona sets up water currents that bring particles within the field of the mouth. The second group rely on capturing prey with the jaws, and the third have the wheel organ, mouth and pharynx combined to form a snare or funnel.

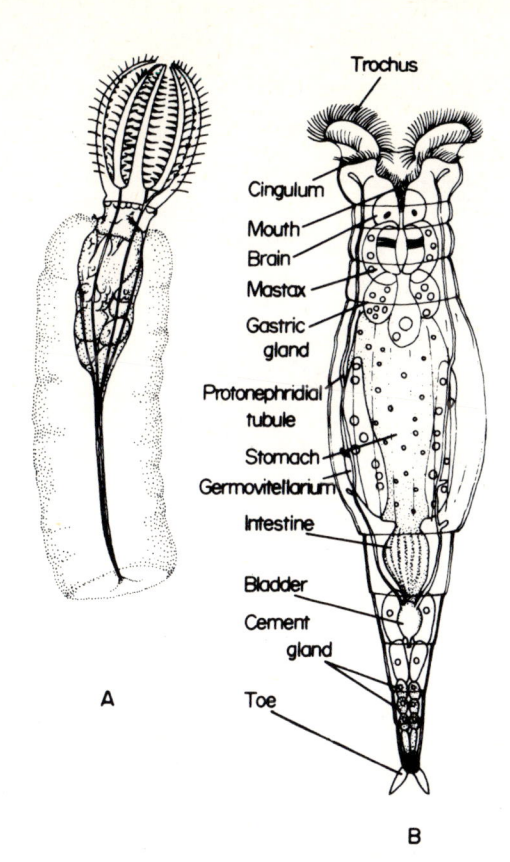

Fig. 11.1 A. *Stephanoceros fimbriatus* ♀, sessile adult (after Jurczyk). B. *Philodina* (after Hickernell).

Movement

Some rotifers are sessile and in these animals only withdrawal and bending are possible. Amongst more active forms some progress with leech-like movements whilst the majority swim rapidly by the action of the ciliated wheel organ.

Co-ordination

a *i* The nervous system is composed of a brain supra-oesophageal in position, from which nerves pass in all directions. Ganglia are associated with the mastax and the foot. There are a number of sense organs serving tactile, mechanical and photic modalities.

ii Nothing is known of nervous physiology.

b Hormones are not known.

Osmoregulation/excretion

a The nephridial system consists of a pair of canals starting in the head region and passing along the body to empty into a posterior bladder. At intervals along their length flame cells or protonephridia are attached.

b The bladder empties periodically, discharging fluid, and is presumably implicated in both excretion and osmoregulation.

Respiration

a There are no special respiratory organs.

b Gaseous exchange must occur across general body surfaces.

Circulation/coelom

There is no circulatory system. The body cavity is a pseudocoel and is fluid-filled. Metabolites and gaseous material may be dissolved in the fluid.

Reproduction

a The sexes are separate, but the larger individuals are all female. These possess an ovary, yolk gland and oviduct. Shelled yolky eggs are shed. The males are small and simple in structure; the majority are without gut, mouth and mastax. They have a short life. They lack a bladder and often an excretory system, but the sperm sac and testes are prominent. There may be a copulatory organ.

b Three forms of reproduction take place in rotifers: (1) mating by separate sexes, (2) asexual parthenogenesis, (3) development of both sexual and asexual generations. In type 1, males impregnate females by entering the body at some convenient point, and releasing sperm into the body cavity fluid. The eggs and the sperm are haploid. Type 2, the Bdelloidea, is exclusively parthenogenetic and males have never been described. Type 3 alternates sexual and asexual forms. The resting egg hatches a female which produces asexually large thin-shelled eggs that give rise to further females. Later in the sequence females begin to lay small eggs that hatch to haploid males which must fertilize other haploid eggs to again produce the resting egg. Such cycles may take one, two or many years to complete.

References

Donner J. 1966. *Rotifers*. Warne & Co. Ltd., London.
Hyman L.H. 1951. *The Invertebrates*, Vol. III. McGraw-Hill, New York.

12 Phylum PRIAPULOIDEA

5 species described. Maximum size a few cm.

Characteristics

1 Marine, benthic.
2 Unsegmented, vermiform.
3 Large body cavity lined by a peritoneum. Still debatable as to whether this represents a true coelom or is to be regarded as a pseudocoel.
4 Eversible anterior end.
5 Nervous system is not separated from the epidermis. There are no ganglia.
6 There are protonephridia containing large numbers of solenocytes, which open with gonads by urinogenital ducts to exterior.

Larval form

Remarkably like rotifers in appearance, with a lorica or chitinous tunic (Fig. 12.1).

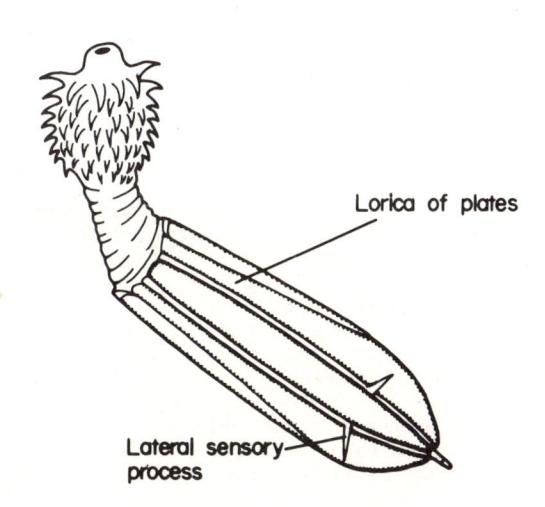

Fig. 12.1 Larva of *Priapulus* (after Lang).

Metamorphosis

Adult status is achieved by a series of cuticular moults.

Adult body form e.g. *Priapulus, Halicryptus*.

The adult is unsegmented, but bears superficial annulations that give the appearance of segmentation. The surface is spinous and bears papillae. There is a large, anterior, retractable proboscis which carries many teeth arranged in concentric, double pentagons. Fig. 12.2 shows the body form of *Priapulus*. Fig. 12.3 indicates the major internal features. A caudal appendage (not in *Halicryptus*) extends posteriorly; its function is not known.

Feeding

Priapulids are carnivorous, capturing prey in the muddy substrate with the introvert (proboscis) and swallowing them whole.

Osmoregulation/excretion

a On each side of the alimentary canal there is an elongate structure within which is situated a protonephridial tubule. To one side of this tubule there are connected many flame cells (solenocytes).

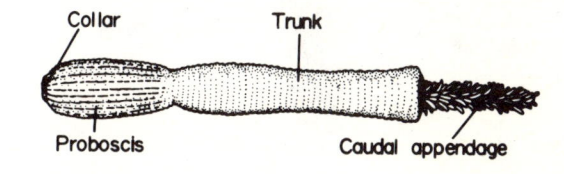

Fig. 12.2 External appearance of *Priapulus caudatus* (after Hammond).

The tube opens through a nephridiopore at the end of the trunk.

b Little is known of the physiology of these organs.

Movement

a Priapulids are typically burrowing animals immersing themselves in muddy substrates.

b The burrowing is accomplished by eversion and retraction of the proboscis under the action of circular and longitudinal muscles of the body wall and the use of the rear portion of the body as an anchor (Fig. 12.4).

c Nothing is known of the muscle physiology.

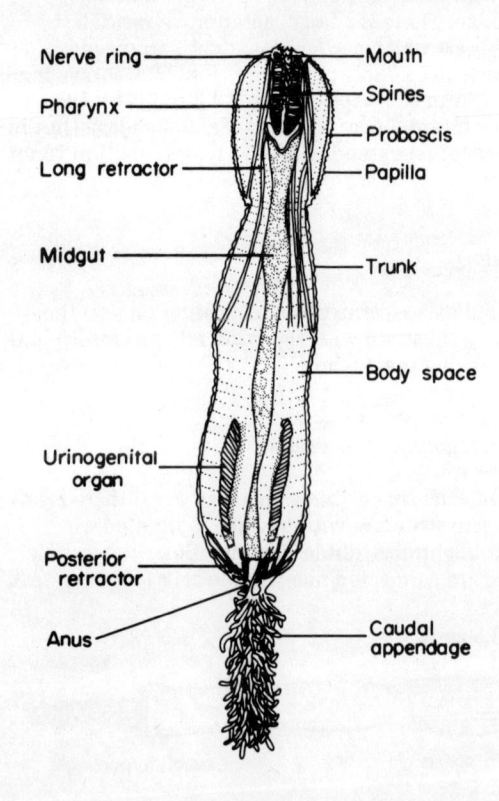

Fig. 12.3 Internal anatomy of *P. caudatus* (after Hammond).

Co-ordination

a The nervous system is simple, consisting of a nerve ring anteriorly around the pharynx, and a mid-ventral cord which is not obviously ganglionated. The nerve cord is located within the epidermis of the animal.

Sense organs may be located on the papillae, and be concerned with mechano- and chemoreception. No eyes are known.

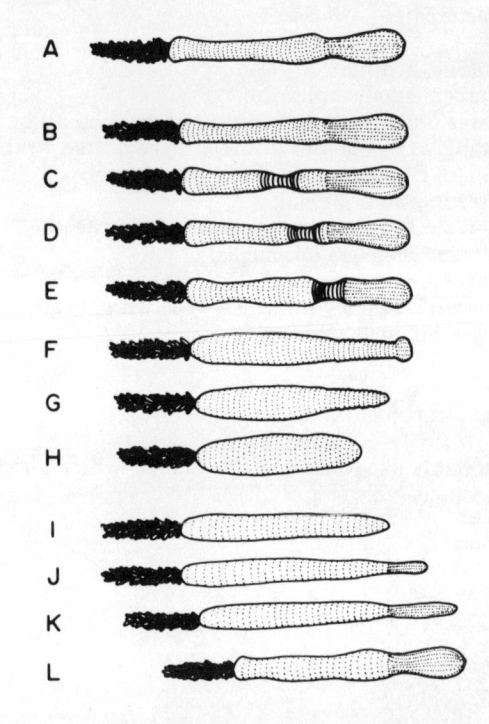

Fig. 12.4 The locomotion of *Priapulus*. A. Animal stationary in the substratum. B. The musculature of the trunk wall beginning to contract. C—E. A contracted zone appearing in the trunk wall and beginning to travel anteriorly. F. The proboscis almost entirely incorporated into the contracted zone. G. The proboscis being withdrawn by its retractor muscles. H. The proboscis fully invaginated. I. The trunk elongating and its anterior part being thrust into the substratum. J. The proboscis being evaginated and forced forward into the substratum. K. The proboscis becoming anchored in the substratum. The trunk shortening and moving forward. L. Completion of cycle (A—L after Hammond).

Phylum PRIAPULOIDEA

b Hormonal co-ordination is not known.

Respiration

There are no special structures for respiration although the caudal appendages have been suggested as sites of gaseous exchange.

Circulation/coelom

a There is no blood system as such. The body cavity has been classified as a haemocoelic space. Recent work on the nature of the bounding membrane of this space seems to indicate that it may be a true coelom. Embryological studies could settle this point.

b The fluid of the coelom is circulated through the space by the body wall movements, and is of great significance as the hydrostatic skeleton.

Reproduction

a The sexes are separate. The gonads are tubular and located alongside the protonephridial tubules into which they open.

b Fertilization is external. Cleavage is total, equal and radial in type.

Phylogenetic relationships

There is some discussion over the exact status of *Priapulus* and its relatives (*Halicryptus*). This depends partly upon the nature of the body cavity and whether or not it is a true coelom. There are some resemblances with Rotifera and with Kinorhyncha and Gastrotricha.

References

Hammond R.A. 1970. The surface of *Priapulus caudatus* (Lamarck, 1816). *Z. Morph. Tiere*, **68**, 255—268.

Hammond R.A. 1970. The burrowing of *Priapulus caudatus. J. Zool. Lond.* **162**, 469—480.

13 Phylum NEMATOMORPHA

Long thin wormlike animals (hair worms), up to 1 m long. About 250 species known.

Characteristics

1 Adults are free-living in fresh water; juveniles are parasitic in arthropods. One genus is marine.
2 Larva is known as the gordioid.
3 Unsegmented, filiform worms.
4 Gut non-functional in adult.
5 Excretory system absent.
6 Pseudocoelomate.
7 Sexes separate, each with a cloaca.

Larval form

The larva (Fig. 13.1) which hatches from the egg is the gordioid and has a brief free-living phase. It is able to penetrate small aquatic animals, but develops to a juvenile only in an insect, centipede or millipede. In other species it may encapsulate. The larva may be ingested in food, and then, using its armed proboscis, bore its way out of the gut and into the haemocoel of the host. Nutrition is via the body surface, the gut being non-functional.

Metamorphosis

In the arthropod host the nematomorph gradually increases in size. The transition to a juvenile occurs without drastic rearrangements but the larval proboscis and muscles degenerate, whilst the cuticle and epidermis are retained to form adult structures. New organs such as the brain appear. The larval gut, if present, elongates but is not open anteriorly in most species. The juvenile eventually leaves the host, and moults to the mature adult. This only occurs when the host is in a moist situation. The host may undergo parasitic castration.

Adult body form e.g. *Nectonema*, *Gordius*.

The adult is very long and thin with females usually longer than males. The mouth is ventral or terminal anteriorly, the anus (cloaca) is ventral or terminal posteriorly.

There is a thick cuticle sometimes adorned with bristles, bumps or warts. Some thickened portions are known as the areoles. The cuticle is fibrous internally and overlies the epidermis, which is single-layered and may form longitudinal cords of which one incorporates the nerve cord. Musculature is arranged longitudinally only. There is a pseudocoel which is almost filled with mesenchyme cells, but some fluid space remains. The skeleton is cuticular and hydrostatic.

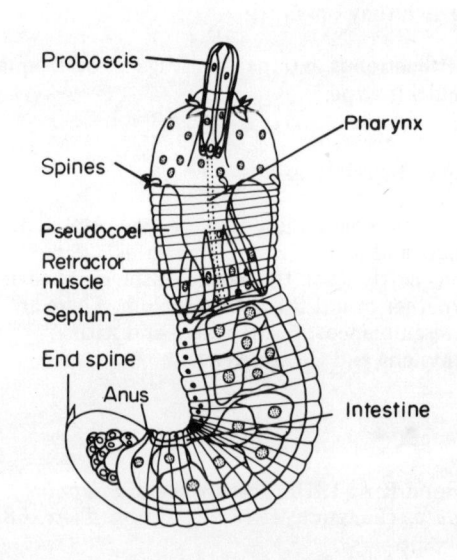

Fig. 13.1 The larva of *Gordius aquaticus* (after Mühldorf).

Feeding

a Feeding does not occur as such in nematomorphs. Absorption of nutrient, even in larvae and juveniles, takes place through the body wall. External secretion of digestive enzymes has been suggested in the early stages.

b The gut does not function in the adult and is at least partly degenerate in all forms. Many have no mouth, a non-patent pharynx, and an epithelial intestine which receives genital ducts towards the rear and opens via a cuticle-lined cloaca.

Osmoregulation/excretion

a There are no excretory organs, but the posterior part of the gut may be involved.

b Nematomorphs inhabit fresh water (except *Nectonema*) and presumably face osmotic problems, but no information on regulation is available.

Movement

a Males are more active than females. They may swim, undulate or coil up (the **Gordian** knot). Aggregations can occur with males seeking out females.

b Longitudinal muscles only are found, functioning against the resistance of the cuticle and hydrostatic contents.

c Physiology is unknown.

Co-ordination

a *i* The nervous system is confluent with the epidermis. There is an anterior cerebral ganglion and a ventral cord, which lies deep to the body wall. Ganglia are not described but nerves serve various parts of the body. Sense organs are simple in structure and may be represented by bristles and spines.

ii No physiology is known but behavioural observations suggest that chemoreceptors are present, since males seek out females for mating, and that hygroreceptors are present, since larvae hatch from their hosts in response to damp conditions.

b No hormones are known, but the evidence reviewed above suggests release of a female pheromone or sex attractant.

Respiration

a There are no special respiratory organs.

b Gaseous exchange presumably occurs by diffusion across the body wall.

Circulation/coelom

a The body space is a pseudocoel, largely occluded by mesenchyme. A small anterior part is cut off by a septum near the brain.

b Any fluid movement must occur during body movements.

Reproduction

a The sexes are separate. The gonads lie in spaces bounded by mesenchyme. Each sex has one pair of gonads. In the male the testes are long, cylindrical organs, each with a sperm duct opening via a penis

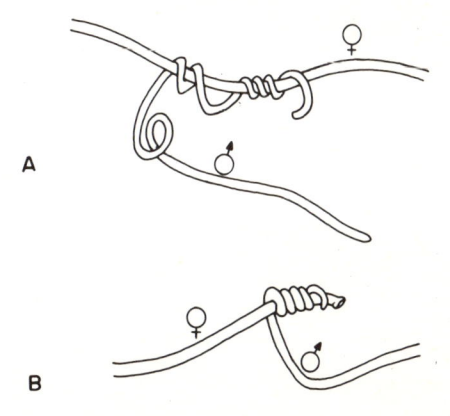

Fig. 13.2 Adults of *Nectonema* in copulation (after Huus).

at the cloaca. The females have long ovaries with oviducts that fuse to a common bladder that opens into the cloaca.

b The sexes copulate, the males seeking out females and coiling their posterior ends around them (Fig. 13.2). Sperm pass into the seminal receptacle adjacent to the uterus, and fertilization occurs internally.

Reference

Grassé P.P. (Editor), 1965. *Traité de Zoologie*, Vol. IV (3). Masson et Cie, Paris.

14 Phylum ACANTHOCEPHALA

Worms ranging in length from 1.5 mm to about 50 cm. About 500 species are described.

Characteristics

1 All are endoparasites.
2 The larvae develop in the ♀ uterus and need further development in a secondary host.
3 The anterior extremity is formed of a protrusible proboscis armed with hooks.
4 There is no gut.
5 There is no coelom, but a pseudocoel is present.

Larval form

There are successive larval types: the acanthor, the acanthella and the juvenile.

Metamorphosis

The first larva is retained within the parent uterus and is held within a shell (hard in worms with terrestrial intermediate hosts, soft in those with aquatic hosts). This is the acanthor, with 3 pairs of larval hooks, and spines on the body surface. It undergoes further development in an intermediate host (which may be an insect, amphipod or isopod). It is then released from the egg and penetrates the host gut wall. The acanthor larval structures regress, and the larva enters the host haemocoel where it starts to elongate as the acanthella. Eventually it reaches a juvenile form with all adult characteristics except the mature reproductive system.

Adult body form e.g. *Acanthocephalus, Moniliformis.*

As for rotifers and nematodes, the cell numbers are constant in Acanthocephala. The worms are elongate, cylindrical or slightly flattened, with a pseudocoel. Females are usually larger than males. The body is sometimes annulated or superficially segmented, and is divided into a presoma (including the proboscis, armed with hooks, and the neck, usually smooth with no hooks) and a long trunk (which may be spinous, or not) (Fig. 14.1).

The proboscis is the diagnostic feature of the group. It bears recurved hooks and is invaginable into a muscular sac or receptacle. The hooks provide purchase on the tissues of the host. The body is covered with a thin cuticle that overlies a syncytial epidermis, which is fibrous. The number of nuclei in the epidermis is very small (6—20). The inner layer of the epidermis carries the lacunar system which is a set of channels and is fluid-filled. Muscles are thin and syncytial.

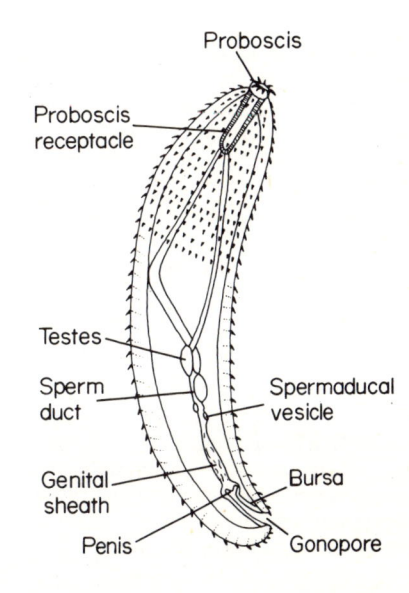

Fig. 14.1 Anatomy of *Acanthogyrus* (after Thapur).

Feeding

a Absorption of nutrient takes place directly through the body wall from the gut of the host.

b There is no gut; the lacunar canals may be concerned in food absorption.

Osmoregulation/excretion

a 1 pair of protonephridia is present with a large number of attached flame cells. The two canals open into the gonoduct and thence to the exterior by a posterior gonopore.

b Physiology is poorly understood; the worms appear to be osmotic conformers.

Movement

a Movement is limited in adult worms that are firmly attached to the host gut lining. Body length when elongated is only 40—50% greater than when contracted. Some trunk bending occurs. Juveniles bore actively.

b The movements shown are due to interactions of body wall muscle and the internal hydrostatic pressure. Proboscis position is governed by the trunk internal pressure, eversion by the separate hydraulic system of the proboscis receptaculum. Fluid from the lemniscal sacs moves to the proboscis and distends it, but does not bring about evagination (Fig. 14.2). Invagination is by retractor muscles.

c Muscle physiology is not known.

Co-ordination

a *i* The nervous system consists of a cerebral ganglion in the ventral wall of the proboscis receptacle with a fixed number of cells, from which a number of nerve trunks arise supplying the proboscis and trunk. Males have 2 genital ganglia. Sense organs are few, there being 3 on the proboscis and a few in the male bursa and penis.

ii The physiology is not known. Sense organs may be mechano- or chemoreceptors.

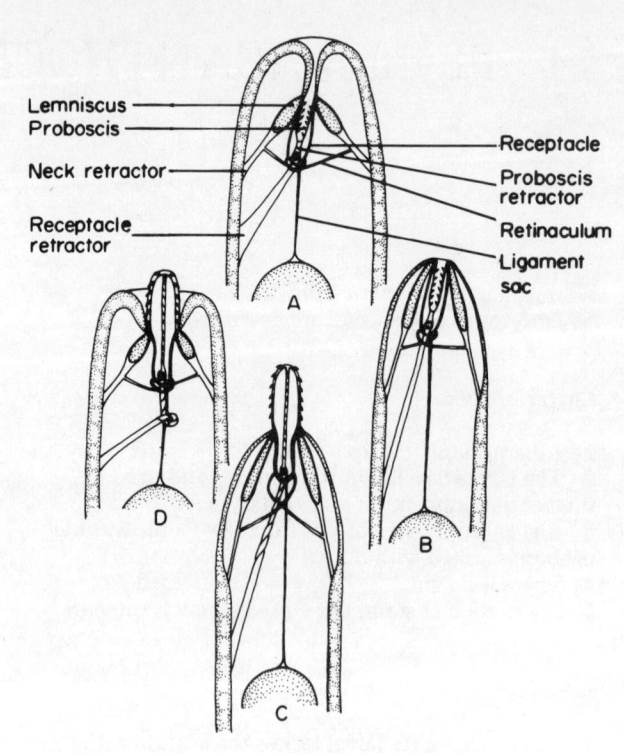

Fig. 14.2 Operation of the proboscis of *Acanthocephalus*. A. Proboscis invaginated, proboscis apparatus fully withdrawn. B. Trunk maximally extended. C. Trunk extended, proboscis everted. D. Proboscis everted, neck retractor muscles contracted (A—D after Hammond).

b No hormones are known.

Respiration

a No respiratory organs exist.

b Anaerobic glycolysis provides most of the metabolic energy of the worms.

Circulation/coelom

a There is a pseudocoel, small in species with two ligament sacs, large in those with only one. This unlined cavity lies between the body wall and the ligament sacs, hollow tubes of connective tissue which run the length of the body and enclose the reproductive organs. The lacunar system, with its several channels, is restricted to the epidermis, not connected to the pseudocoel, and is fluid-filled.

b   Any movement of fluid within the body is solely due to body wall movement.

Reproduction

a   The sexes are separate. Males possess two testes in the ligament sac. Each has a sperm duct that passes caudally (in some species provided with cement glands). The two ducts fuse to form a urino-genital tract with the protonephridial canal, leading to the penis which is protruded. The penis lies in a cavity, the bursa, that is in turn eversible to the exterior for grasping the female in copulation. Females possess 1 or 2 ovaries that break up into ovarian balls floating in the ligament sac or pseudocoel.

b   The eggs are fertilized in the pseudocoel as they reach the surface of the ovarian ball, and develop into larvae within the pseudocoel. They are then taken up by the muscular uterine bell (first part of the female canal), move down the uterus (long and muscular) and are voided at the vagina.

Life-cycles

The Acanthocephala are all parasitic as adults in vertebrates. They infest fish, birds, reptiles and mammals. They show enormous reproductive capacity and complex life-cycles. All require one intermediate host, but some have two with the latter acting as a transporting medium. Small invertebrates, particularly arthropods, are favoured intermediates.

References

Crompton D.W.T. 1970. *An Ecological Approach to Acanthocephalan Physiology* (Cambridge Monographs in Experimental Biology). (C.U.P.).

Nicholas W.N. 1967. The Biology of the Acanthocephala, in *Advances in Parasitology* (ed. B. Dawes) Vol. 5. Academic Press, London and New York.

15 Phylum NEMATODA

Roundworms vary from minute to massive (2 m long). Estimates of numbers vary greatly, with an upper figure of over half a million species.

Characteristics

1 Representative species are aquatic (both marine and fresh-water), terrestrial or parasitic.
2 They are vermiform, usually cylindrical in shape, triploblastic and non-segmented.
3 Covered with cuticle, having no external cilia.
4 Epidermis composed of four or more longitudinal cords.
5 Muscle fibres arranged only longitudinally.
6 There is a fixed cell number.
7 Pharynx long and triradiate.
8 Excretory system of 1 or 2 gland cells or canals.
9 Male duct opens into rectum, female has gonopores.

Larval form

Nematodes hatch as juveniles. The term 'larva' is not appropriate although sometimes used.

Metamorphosis

There is no metamorphosis. The juveniles develop by a series of moults during which the entire cuticle is lost, and at each stage the reproductive organs increase in complexity. Maturity is achieved at the fourth moult.

Adult body form e.g. *Ascaris, Ancylostoma, Heterakis.*

Nematodes are all elongate, cylindrical animals with either fusiform or filiform shape. Modifications of anterior or posterior extremities are noted in many species. There are no obvious body regions (Fig. 15.1).

The mouth lies anteriorly, with six labia around it; further rings of such papillae may appear, making concentric arrangements. This imparts a biradial symmetry to the anterior region. Various sense organs are located in this area, and other cuticular specializations are found outwith the labial portion. The major part of the body surface, however, is smooth but there are many possible protuberances and modifications in the way of papillae, spines, ridges etc. In free-living worms there is a caudal (rectal) gland. The body wall is composed of cuticle (Fig. 15.2), epidermis (hypodermis) and a muscle layer. The cuticle is moulted periodically to allow growth. Internally there is a cavity around the gut and gonads. This is a pseudocoel not lined by mesoderm cells. Only longitudinal muscle fibres exist (Fig. 15.3). The gut is not muscularized. The skeleton is hydrostatic and cuticular.

Feeding

a The adopted feeding method is a reflection of the habitat and mode of life of the nematode. Free-living worms are carnivorous, herbivorous or saprophagous with a diet of small invertebrates, diatoms or algae, or dead and decaying organic substances respectively. Parasitic nematodes may infect either plants or animals. Almost all animals appear to have nematode parasites and for man and domestic animals there may be serious consequences. Blood, gut, lungs and other organs may all harbour nematode worms.

b Although the diet is varied the feeding organs are comparatively uniform throughout the group. The pharynx is suctorial, and this coupled with mobile labia and spines or teeth, provides the ingestion mechanism. Internal parasites are often firmly attached to the host tissue, and may rasp or suck the food into the mouth. Autolysis of host tissue

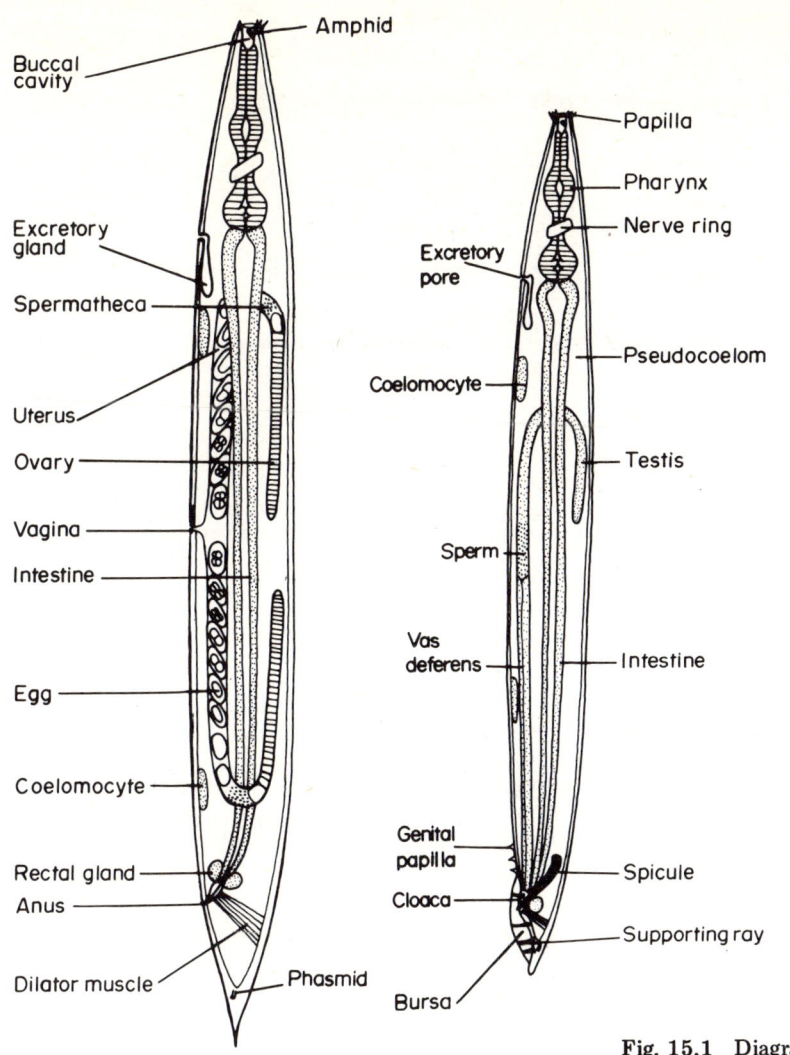

Fig. 15.1 Diagram of internal anatomy of male and female nematodes (after Lee).

may occur. Some nutrients pass directly across the body wall.

Osmoregulation/excretion

a   Glandular organs are seen in some marine nematodes. A single, large, gland cell with a neck opens ventrally at an excretory pore, far anteriorly. This is termed the renette. More complicated renette plans are found in some fresh-water and terrestrial forms (Fig. 15.4). This simple cellular organ develops in more advanced forms into a pair of canals linked transversely, in some species with renette cells in continuity but otherwise with no excretory cells attached. The canals represent intracellular channels.

Phylum NEMATODA   67

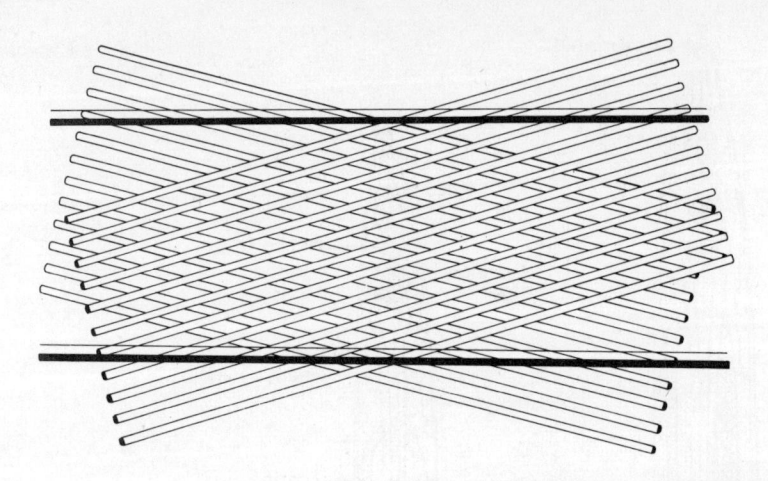

Fig. 15.2 Fibrous layers of the cuticle are arranged in right and left-handed spirals that cross one another at angles varying with degree of elongation of the animal (after Harris and Crofton).

b The physiology of ionic regulation, water clearance and nitrogenous excretion is not well known. Osmoregulation is carried out by many species. It seems probable that the body wall and gut may both be implicated, whilst the canals may be relatively unimportant. Filtration into the canals under high pseudocoelomic hydrostatic pressure has been described. Ammonia, urea and uric acid have been reported as nitrogenous waste materials.

Movement

a Nematodes retain their typical vermiform, cylindrical shape throughout all cycles of movement. Most movement is stereotyped flexing of the body, usually with little directional component. Swimming is possible for a few species, crawling occurs (sometimes with sinuous contortion, sometimes with rather angular progression), and attachment by temporary anchors (bristles, secretions) may allow leech-like movement or a form of walking. Rapid lashing in a dorso-ventral fashion is very obvious in many small nematodes. Parasitic worms exhibit little movement once the adult habitat is attained.

b Nematodes possess only longitudinal muscles. All movement is the result of interactions between these muscles, the cuticle and the internal fluid. Muscles exert forces during contraction and shortening, and do not actively extend. Extension is brought about by the redistribution of hydrostatic forces.

c The muscles of nematodes have 'fingers' that are non-contractile and which abut onto the nervous system. These muscle tails act as conduction pathways to the contractile mechanism.

The mechanism of movement in nematodes is as follows: there is a strong cuticle and a high internal hydrostatic pressure (up to 400 mm Hg). Contraction of muscle on one side of the body brings shortening, and an increase in hydrostatic pressure; relaxation is followed by lengthening as the high internal pressure extends the muscle. This is repeated alternately from side to side. The cuticle does not rupture under such conditions, and mechanical strength is given by a helically-arranged fibrous collagen layer.

Co-ordination

a *i* The nervous system, for *Ascaris*, has been described in great detail, and most nematodes have the same basic scheme, which is a circum-oesophageal ring around the pharynx, and ganglia (paired lateral and one or two ventral). From this ring arise nerves supplying the anterior labia or papillae, and 6 posterior-going longitudinal nerves, with ganglionic modifications at the anus, cloaca and other special regions. Sense organs are plentiful, with a variety of sensilla types (hairs, bristles, warts). Some of these contain ciliary elements, the only signs of cilia in nematodes. Amphids are a particular kind of organ, consisting of a small dimple in the cuticle anteriorly, and may be chemosensory. They are most obvious in free-living forms. Phasmids are glandular-sensory, open near the tail and are best developed in parasitic species. Eyes are found in free-living species.

ii Physiology is not well known, but work is in progress at present on the nervous system of *Ascaris*.

b Neurosecretory cells in cephalic ganglia show cyclic activity related to moulting. Hormones may control production of moulting fluid enzymes.

Respiration

a There are no respiratory organs.

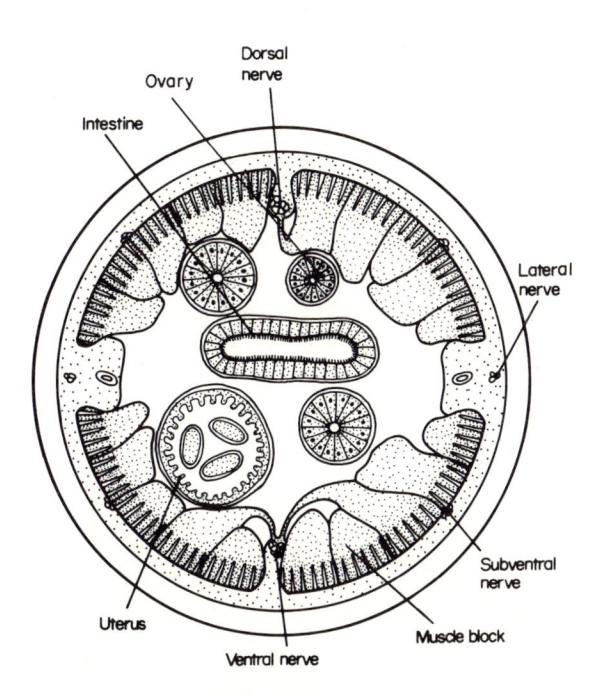

Fig. 15.3 T.S. through mid-region of a female nematode to show arrangement of longitudinal muscles (after Lee).

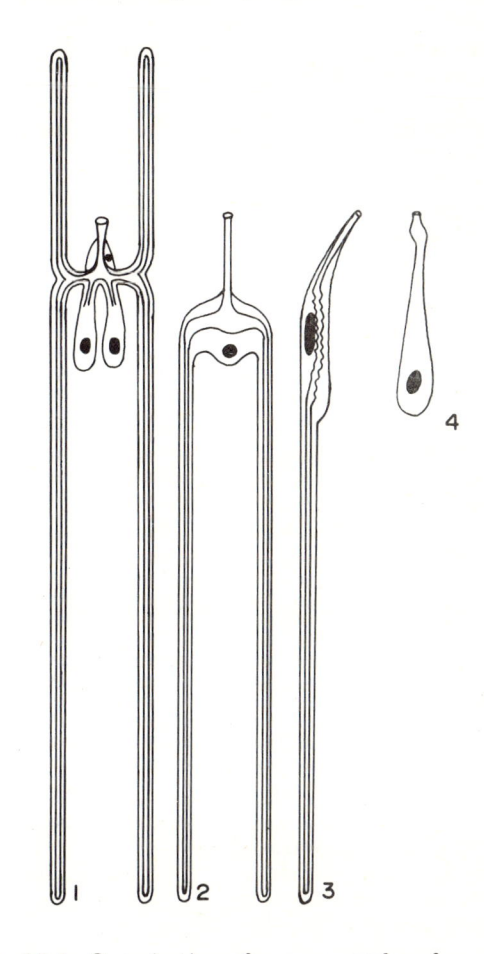

Fig. 15.4 Organization of some examples of excretory systems. (1) Rhabditoid, (2) Cephaloboid, (3) Anisakid, (4) Chromadorina, Monohysterina and Enoploidea types (after Chitwood & Chitwood).

Phylum NEMATODA

b Free-living nematodes must obtain their oxygen requirements by diffusion from the environment. Zooparasitic worms are facultative anaerobes, obtaining energy from glycogen but requiring oxygen for some purposes.

Circulation/coelom

a There is no circulatory system and no coelom. The fluid-filled space between body wall and gut is a pseudocoel. Some large cells produce extensions and membranes that cross the pseudocoel. These are fixed, non-amoeboid and not phagocytic. They may be important metabolically.

b Fluid movement must occur by movements of body wall muscles.

c Haemoglobin is present in the pseudocoel of some nematodes.

Reproduction

a The sexes are usually separate although hermaphroditism occurs in some terrestrial species. The female possesses 2 ovaries, 2 oviducts and 2 uteri joining at a single vagina and opening at a gonopore in the anterior third of the body. Males are smaller, are curved posteriorly, and possess bursae and other copulatory devices. The testis is single and there is a seminal vesicle, leading to the exterior via an ejaculatory duct located alongside the anus. Setae and spicules are found in males at the cloaca.

b Fertilization is internal. The sperm are amoeboid, and progress to the seminal vesicle after copulation. Shelled (chitinous) eggs are laid and development begins immediately; in some species juvenile worms are present in the eggs before they are shed, and in some ovoviviparity occurs and juvenile worms hatch within the uterus. Parthenogenesis also occurs amongst terrestrial species.

The further development of parasitic nematodes is extremely variable and complex and worthy of longer description than can be given here.

References

Grassé P.P. (Editor) 1965. *Traité de Zoologie*, Vol. IV (3). Masson et Cie, Paris.

Lee D.L. 1965. *The Physiology of Nematodes* (University Reviews in Biology). Oliver and Boyd, Edinburgh and London.

16 Phylum ENTOPROCTA

Entoprocts account for about 60 species, all very small (5 mm or less).

Characteristics

1 All are marine except *Urnatella* which lives in fresh water.
2 They are all sessile, either solitary or colonial.
3 They possess a circlet of ciliated tentacles.
4 The group is non-coelomate, although possessed of a pseudocoel of gelatinous mesenchyme cells.
5 The gut is U-shaped with both mouth and anus opening within the tentacular circle.
6 Entoprocts possess protonephridia.

Larval form

The larva of *Pedicellina*, which is best known, is called a trochophore although it differs considerably from annelidan trochophores. The larva is very mobile and variable in shape. *Loxosoma*, a primitive entoproct, has a larva of completely different form (Fig. 16.1).

Metamorphosis

Entoproct larvae have a short free-ranging existence after which they settle and attach to the substrate. Some have longer, pelagic lives. A pedal gland produces secretion that aids in attachment. The larva undergoes considerable alteration to form the adult. The interior of the larva undergoes a form of torsion with the rotation of the calyx through 180°. All larval parts persist in the adult except the apical organ, the pre-oral organ and the ciliary girdle.

Adult body form e.g. *Pedicellina*, *Loxosoma*.

Entoprocts are small, either solitary or colonial animals. They are generally attached to solid substrates or epizoic on other animals. Their resemblance to hydroids is superficial and they can be easily distinguished by the ciliation of the tentacles. The major portions of the body are the tentacles (arranged as a crown), the calyx (the body region that carries the crown) (Fig. 16.2), the stalk (carrying the calyx) and the basal attachment.

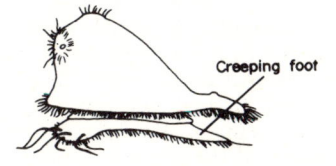

Fig. 16.1 Larva of *Loxosoma* (after Jägersten).

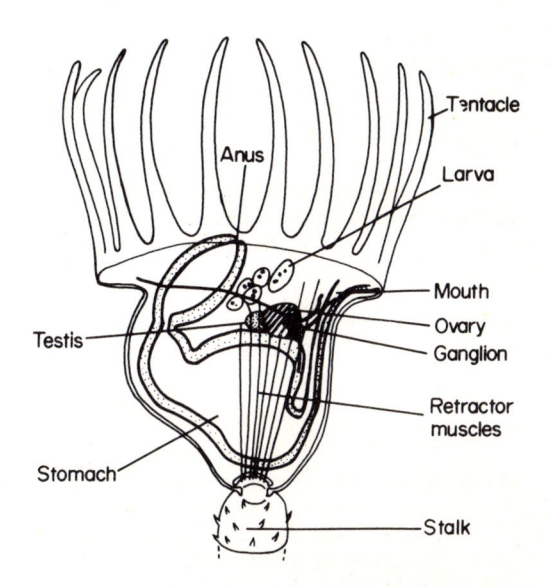

Fig. 16.2 The calyx of *Pedicellina* demonstrating diagrammatically the major organ systems (after Atkins).

The tentacles are usually all of similar length and are ciliated on their inner faces. The mouth receives material from the tentacles, and both it and the anus are located within the ring. The anus may be mounted on a small conical projection. The stalk is an outgrowth of the calyx and in most species is partially separated from it by a septum. It takes many forms, being either smooth or roughened, cylindrical or bead-like, muscular or rigid. The stalk base may anastomose and branch to form ramifications over the substrate, thus giving rise to a colony.

The structure of the body wall is similar throughout the colony. It is cuticulate. The larger part of the body volume is taken up by the pseudocoel which is a gelatinous material containing mesenchyme cells. Some free amoeboid cells are present.

Feeding

a Entoprocts are ciliary suspension feeders on diatoms and other small organisms.

b The tentacles provide a large feeding surface. Water is moved directionally from outside to inside the crown and ejection is vertically above the animal. Particles are removed from the water current. The gut is recurved and mouth and anus open close together.

Osmoregulation/excretion

a A pair of protonephridia is found in each calyx ventral to the stomach and between the oesophagus and the ganglion. The two ducts coalesce and open at a single nephridiopore.

b Nothing is known of the excretory or osmoregulatory capabilities of these organs.

Movement

a Movements of entoprocts are strictly limited. Only the solitary *Loxosoma* is able to move freely, which it does in looping steps rather like a leech. Colonial species are anchored to one spot but can, in long-stalked varieties, sway from side to side, and in all species the delicate tentacular crown can be quickly withdrawn upon tactile stimulation.

b Apart from the basic anatomy nothing is known of the muscle of entoprocts.

Co-ordination

a *i* There is only one major ganglion in entoprocts, lying ventral to the stomach. From this ganglion pairs of nerves innervate the tentacles, each terminating in a ganglion. A nerve net is reported in the stalk, stolons and muscular areas of *Barentsia*. Sense organs are found at the tentacle tips as sensory bristles. In the active *Loxosoma* group a pair of oral sense organs are reminiscent of rotifer antennae.

ii Nervous physiology is unstudied in the group but it is known that slight tactile stimulation of the sensory bristles leads to withdrawal of the tentacles.

b No hormones are known.

Respiration

a Respiration is probably a general function of the body wall and tentacles.

b Physiology is not known.

Circulation/coelom

As acoelomate animals there is by definition no coelom. Circulation, if any, is by wandering amoebocytes.

Reproduction

a Some entoprocts are hermaphrodite and may be protandrous; in others the sexes are separate. Gonads are one pair, ventral to the ganglion. In some hermaphrodites a pair of testes lies posterior to the ovaries. Gametes are shed via a single gonopore.

b *Sexual reproduction* involves small, yolky eggs which are fertilized in the ovaries (*Pedicellina*). The eggs are invested by a loose membrane that forms a stalk for attachment to the vestibular wall. The eggs are brooded, and gradually pushed peripherally as new additions are made. A succession of stages is found in the brood chamber until free-swimming larvae are released.

c   Entoprocts reproduce abundantly by *asexual* budding. Depending upon species, buds may appear on the calyx or the stalk and stolons. The whole organism develops from the ectoderm and mesoderm. Growth of complete colonies can take place by regeneration from very small remnants and calyces removed are quickly replaced.

Reference

Marshall A.J. & Williams W.D. (Editors) 1972. *A Textbook of Zoology, Invertebrates* (7th Edition of Parker & Haswell, Vol. I). Macmillan, London.

17 Phylum ANNELIDA

There are approximately 8,750 species of annelids. Many are microscopic but a few reach great length, e.g. *Megascolex*, up to about 4 metres.

Characteristics

1 Annelids are to be found in marine, fresh-water and terrestrial habitats.
2 The larva is the trochophore.
3 The body is vermiform, and segmented. Each segment is separated from contiguous ones by a transverse septum, although this basic feature may be modified in some forms.
4 Coelomate, the coelom being mesodermal (schizocoelic) in origin.
5 Possess nephridia and coelomoducts typically, for excretory and reproductive purposes.
6 There is a hydrostatic skeleton, the animal being bounded by a thin, flexible cuticle.
7 May have chaetae, hard, bristle-like structures projecting from the body wall.
8 The group is triploblastic, and has a body wall musculature of two layers (external circular and internal longitudinal muscle).
9 CNS of preoral ganglia linked by connectives to a pair of ventral ganglionated cords.
10 Circulation is closed and tubular.
11 Reproduction may involve copulation; cleavage is spiral and development is determinate.

Larval form

The larval type of the annelids is the trochophore, although this is only found in a number of polychaetes. Oligochaetes usually show direct development, as also do Hirudinea, with juveniles hatching in an immature adult form.

The trochophore (Fig. 17.1) is typically a spherical organism with a major ring of cilia (the prototroch) lying just anterior to the mouth. There are subsidiary groups of cilia on the larval surface, the most important being the anterior apical plate. This is thought to be a sensory area.

Some trochophores, e.g. the mitraria, show modifications of the basic organization. Mobility is still conferred by the prototroch ring of cilia.

Metamorphosis

The adult develops from the larva by an elongation of the post-oral region. The portion anterior to the prototroch remains as the prostomium of the adult worm. The region behind the mouth elongates by the addition of mesodermal blocks progressively, carrying the larval anus to the rear, the gut extending in simple fashion between the larval mouth and the anus.

Metamorphosis is often accompanied by a change in larval behaviour. Newly-hatched larvae are relatively strong swimmers and may migrate to the surface of the sea. They later become negatively phototropic and positively geotropic, moving towards the sea bed. There they complete the metamorphic changes, providing the habitat is suitable, e.g. *Sabellaria* metamorphoses in contact with other adult individuals already forming a reef. The appropriate stimulus is contact with the cement of existing tubes of the colony. The chemical nature of appropriate substrates is precise.

Adult body form

Adult annelid worms are elongate animals, and show metameric segmentation. There is an internal division of the body by regularly repeated septa. Within the segments of the body there is serial repetition of various organ systems. This is especially marked in polychaetes, less so in oligochaetes and almost obscured in leeches.

It will be convenient to discuss the variations of body form within the classification of the group.

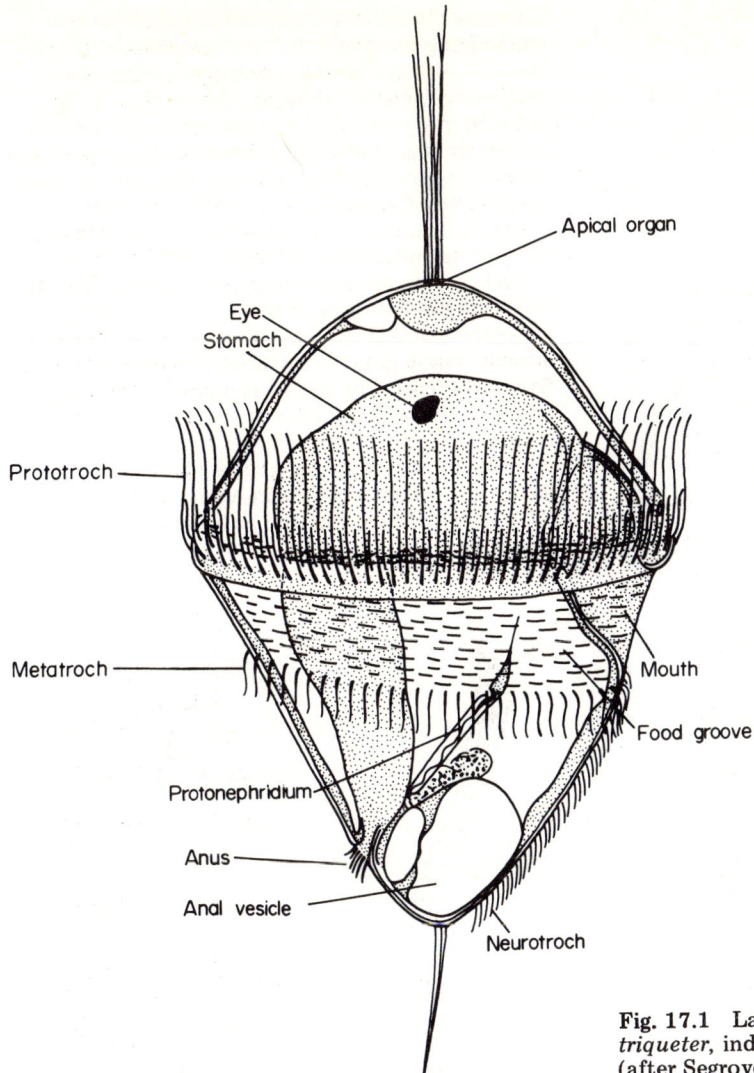

Fig. 17.1 Late trochophore of *Pomatoceros triqueter*, indicating major anatomical features (after Segrove).

1 *Class* POLYCHAETA strongly segmented animals although some genera such as *Arenicola* lose some septae. Many chaetae are present, borne on fleshy outgrowths of the body wall known as parapodia. The chaetae are hard, secreted by ectodermal pits and composed of β chitin tanned by polyphenols. In some species they are siliceous. In swimming forms the parapodia are expanded into paddle shapes.

Polychaetes usually have a distinct head, carrying appendages, jaws and sense organs. This is especially so amongst the errant or free-ranging species (e.g. *Nereis, Phyllodoce*), less marked amongst burrowers such as *Arenicola*, and amongst tubicolous examples like *Sabella* and *Amphitrite* (Fig. 17.2A).

The group is mainly marine, with a few estuarine forms. The animals are dioecious, with gonads repeated along the length of the body, as also are the excretory organs, the nephridia.

The most notable variations upon the body type

Phylum ANNELIDA

extreme. The animal is enclosed totally within a manufactured tube (either of mucopolysaccharide alone or of this material plus incorporated sand grains or calcium carbonate). The body is differentiated into regions. A fan-like crown of tentacles anteriorly is carried on the broad thoracic portion in which notopodia are conical with capillary chaetae, and neuropodia have short, toothed chaetae. Neuropodia are for anchoring purposes, notopodia for moving the animal up and down the tube. To the rear is the slimmer abdomen where the chaetal arrangement may be reversed.

Archiannelids (Fig. 17.3), formerly considered as a small marine group of unusual character, derived from polychaetes by loss of parapodia and chaetae and retention of juvenile characters such as cilia and continuity of the nervous system with the epidermis, are now considered to be polychaetes adapted to a habitat in the interstitial fauna. *Nerilla*, *Saccocirrus*, *Protodrilus* and *Polygordius* are examples.

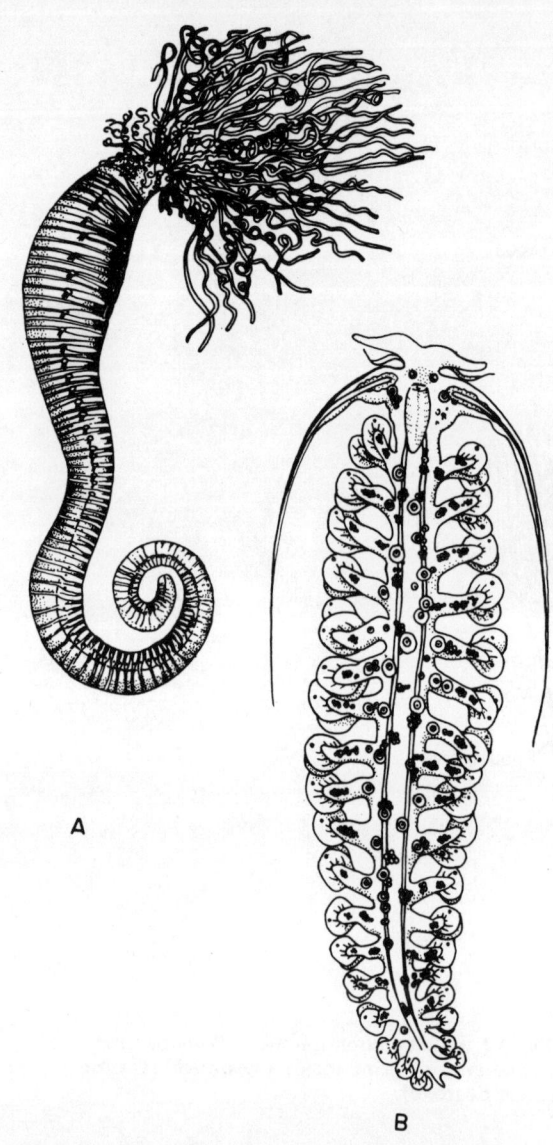

Fig. 17.2 Polychaeta. A. *Amphitrite* (after McIntosh). B. *Tomopteris* (after Greef).

are shown by the burrowers in which the head is reduced, but which possess a proboscis that is utilized in burrowing, and whose coelomic space is large and uninterrupted by septae, enabling shunting of body fluid to and fro during the process. In tubicolous polychaetes modification is more

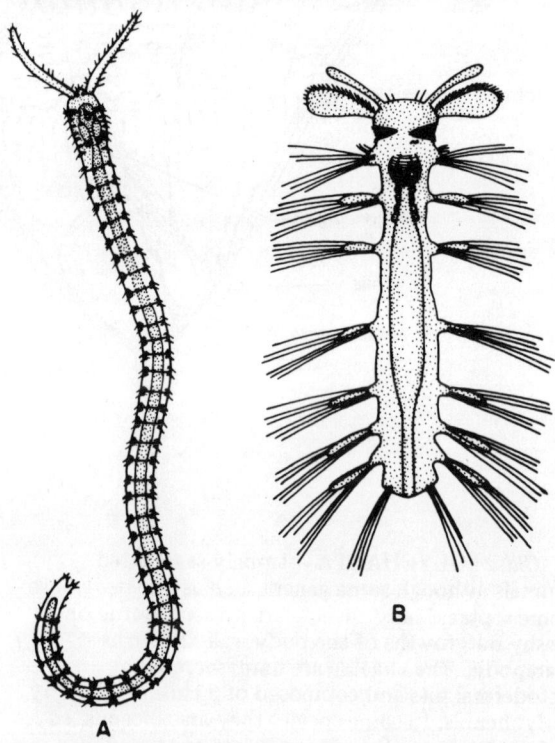

Fig. 17.3 Archiannelida. A. *Protodrilus leuckartis* (after Pierantoni). B. *Nerillidium* (after Remane).

Phylum ANNELIDA

2 *Class* OLIGOCHAETA annelids with few chaetae that are not mounted on parapodia. The prostomium is distinct and there are no appendages anteriorly (Fig. 17.4A). The pharynx is not eversible although it is utilized during burrowing. The group is hermaphrodite with the gonads restricted to certain segments, and passing their contents to the exterior via special genital ducts (coelomoducts); spermathecae and a clitellum are present when mature (Fig. 17.4A,B); copulation and cross-fertilization take place. Habitat may be terrestrial (Terricolae), e.g. *Lumbricus*, *Eisenia*, or fresh-water (Limicolae), e.g. *Tubifex*, *Stylaria*.

The body form is relatively uniform throughout the group even amongst the Megascolecidea which grow to several metres in length. Chaetae are few and occur along the sides of the body, being long and fine in *Stylaria*, short and thick in *Lumbricus*. Otherwise the major external features are the clitellum in mature specimens (this being a modified

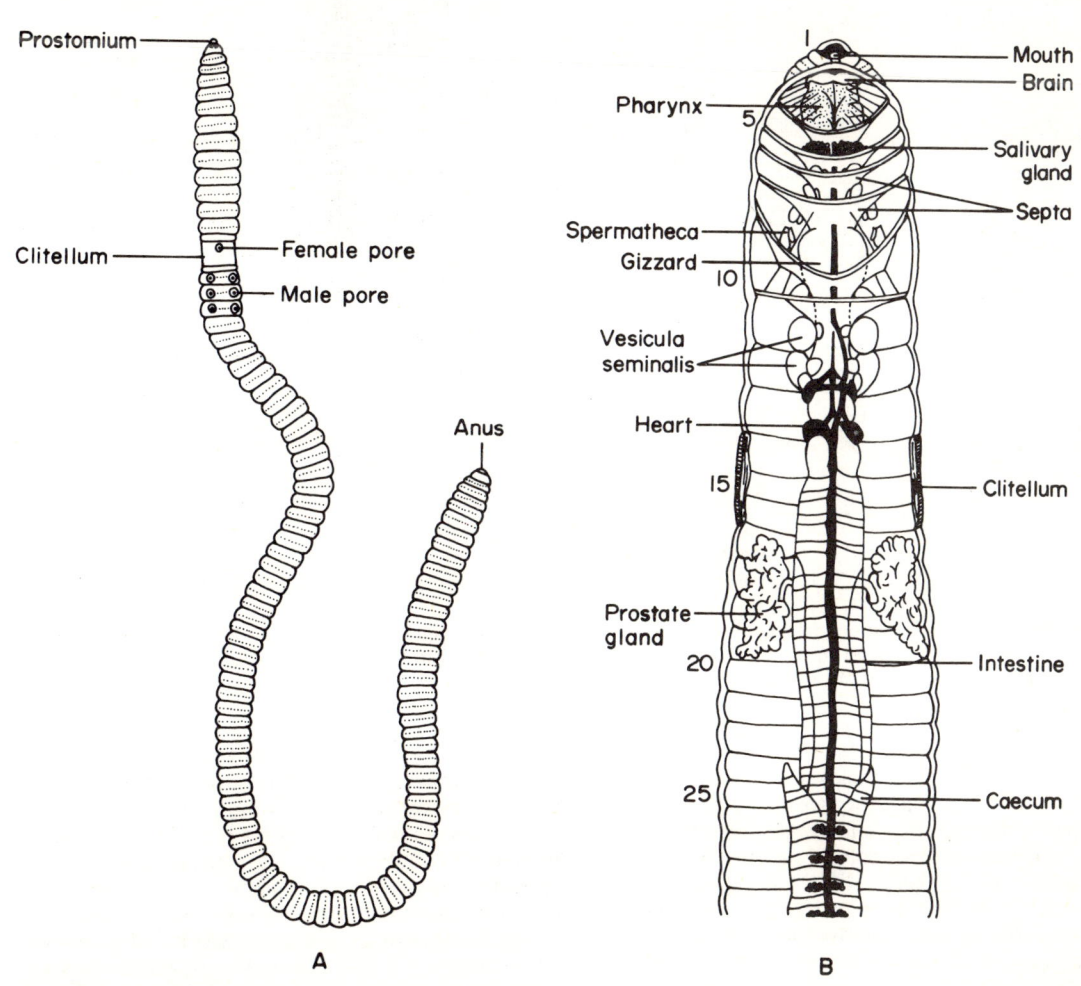

Fig. 17.4 Oligochaeta. A. *Pheretima posthuma* (after Bahl). B. Anatomy of the anterior region of *Pheretima* (after Bahl).

Phylum ANNELIDA   77

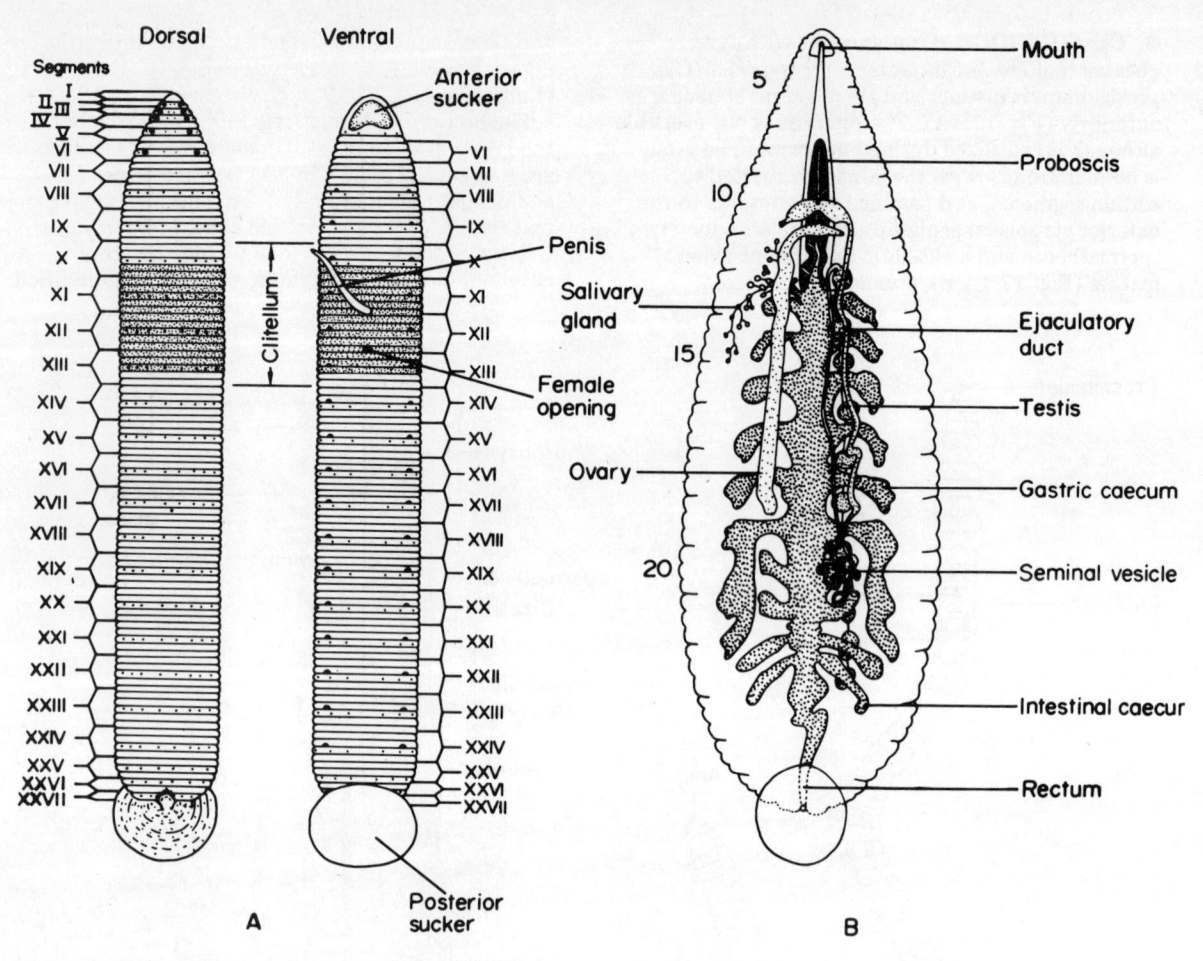

Fig. 17.5 Hirudinea. A. Salient external features of *Haemopis sanguisuga* indicating number of somites, annuli, gonopores, nephridiopores and sensory papillae (after Mann). B. Internal anatomy of a typical leech, *Glossiphonia*, (after Harding & Moore).

region of the epidermis that produces the mucus cocoon in which the eggs are laid), the nephridiopores by which the excretory (osmoregulatory) organs open to the exterior, and dorsal pores in each segment. The gut is notable for the presence of calciferous glands associated with the oesophagus, the function of which is still a matter for debate but which are almost certainly involved in acid-base relations.

3 *Class* HIRUDINEA the leeches. Examples include *Hirudo*, *Haemopis* and *Pontobdella*. This is a short-bodied group with a fixed number of segments (33), but the segmentation is obscured by the development of superficial annuli that give the appearance of greater numbers (Fig. 17.5A). The number of annulations per segment is variable according to species and segment. There are no parapodia and no chaetae (except for *Acanthobdella*). Some segments are modified to form anterior (prostomium + first 2 segments) and posterior (last 7 segments) suckers. The coelom is largely occluded by mesenchyme tissue, known as botryoidal tissue, which leaves only small tubular blood sinuses. The

78 Phylum ANNELIDA

group is hermaphrodite (Fig. 17.5B) and clitellate. Leeches are intermittent ectoparasites.

Acanthobdella peledina (Fig. 17.6) takes an intermediate position between the leeches and the oligochaetes. It is parasitic on salmon. There is no anterior sucker though the posterior sucker is well-formed (composed of only 4 segments). Only 30 segments are represented, each divided by four annuli, and chaetae are found in segments four to six. The coelom is continuous, with only intersegmental septa as in oligochaetes. The testes are tubular and extend through several segments and the gut is simpler than in the remainder of the leeches.

The skeletal apparatus of annelids is hydrostatic in type. The coelomic space is fluid-filled and bounded by the muscular body wall, covered by a flexible cuticle. Pressure changes brought about by contraction of muscles in one part are transmitted rapidly to other areas if the fluid column is continous, as in some burrowers; if segmented, the pressure changes are localized within the segment, and length or diameter changes are imposed, the volume remaining constant (Fig. 17.7).

Feeding

Amongst annelids several feeding styles are found. Many free-ranging polychaetes are predatory and carnivorous, with strongly-developed jaws and proboscis, and associated sensory apparatus.

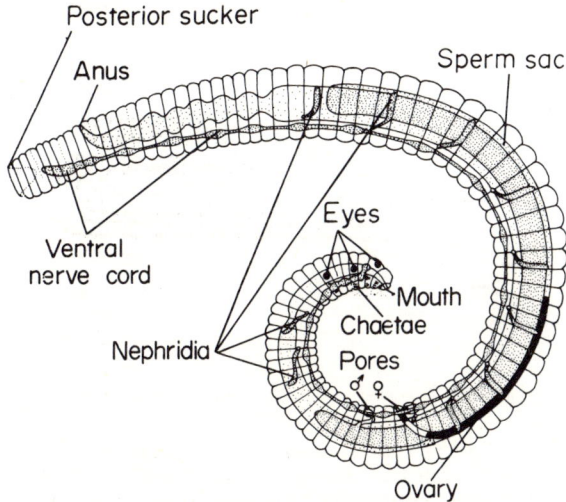

Fig. 17.6 *Acanthobdella*, showing position of chaetae, testes and gut (after Livanoff).

Burrowers may be filter-feeders, drawing water through the burrow and extracting nutrient from it by a variety of mucus net methods. Alternatively, burrowing worms may eat sand and mud and draw nutrients from these by digesting contained fauna and flora. Tubicolous worms are almost all suspension feeders, raising a crown of tentacles above the substrate (support being given by the tube), and passing water across the highly ciliated, tentacular surfaces. Sensory cells are present here, and are probably mainly mechanosensory for initiating protective reflexes. Deposit feeders (e.g. *Terebella*) search wide areas of substrate, continually extending and retracting the very long tentacles of the anterior end. Food is moved in a groove by cilia upon the surface.

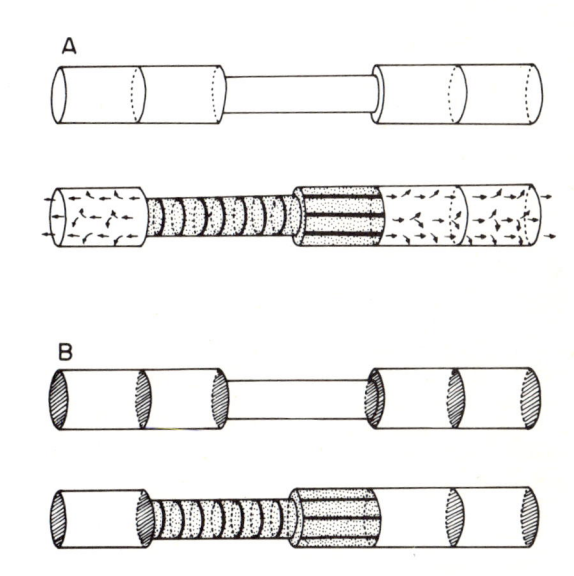

Fig. 17.7 Diagram to illustrate volume and length changes of annelids. The figure shows the essential mechanical difference between non-septate and septate worms during the passage of locomotory waves along the body (after Clark). A. In a non-septate worm, as one region of the body is shortened by contraction of the longitudinal muscles, and an adjacent region is elongated by contraction of the circular muscles, the increase in fluid pressure in these regions (stippled) is dissipated and transmitted to all other parts of the body wall. B. In a septate worm, the pressure changes are ideally localized within the segments undergoing change of length. Sphincters prevent movement of fluid through the apertures around the nerve cords.

Phylum ANNELIDA

Oligochaetes are foraging herbivores, obtaining food (leaves and other vegetable matter) both from the surface of the ground and during the course of burrowing. The gut is adapted for a somewhat cosmopolitan diet and cellulases and chitinases may be among the digestive enzymes present.

Leeches are more specific feeders, the majority being intermittent ectoparasites. They feed occasionally, by suctorial means, after penetration of the host surface by hard teeth in the buccal cavity. Large volumes of blood may be taken in, which remain fluid within the crop, an anticoagulant 'hirudin' being secreted. Digestion is slow and due to the action of bacteria, the capacity for producing proteolytic enzymes having been lost, at least in *Hirudo*. A few leeches and polychaetes are permanent parasites.

Osmoregulation/excretion

a Regulation of ions and of nitrogenous substances may be a function of the general body surface, and of the special segmental organs, the nephridia. In some species (e.g. *Pheretima*) the nephridia open into the rear part of the gut. Other examples have nephridiopores opening on the majority of segments.

b There is evidence that movement of ions such as Na^+, K^+, Cl^-, occurs in both directions across the body wall, and also that some organic substances may be removed from the environment by active uptake mechanisms, especially in marine polychaetes.

Osmoregulation in marine species is a question of preventing influx of ions and removal of water, leading to desiccation. Terrestrial earthworms also face problems of desiccation under certain climatic conditions. Oligochaetes and most leeches are essentially fresh-water animals liable to flooding by massive influx of water.

Limitation of loss of ions or removal of water is a function of the nephridia (Fig. 17.8). These small, ciliated, tubular organs may be numerous or few, open internally to the coelom or closed, voiding to the exterior or to the intestine. Some experimental evidence exists to indicate a production of urine different in composition to body fluids. Leeches may lose up to 80% of total weight and still be revived upon immersion.

Excretion of nitrogen, as ammonia in animals normally immersed in water, or as urea in earthworms subjected to starvation or water deprivation, takes place in unknown fashion. Such substances may derive from the coelom or the blood, and be removed via the skin or the nephridia.

Movement

a Annelid movement is due to the action of the musculature upon the enclosed fluid-filled coelomic space. This may be aided or exaggerated in some cases by parapodial and chaetal development (as in swimming polychaetes) or by suckers (as in hirudineans).

Peristaltic waves pass along the body, providing alternately elongating or shortening portions. At the same time the body may be thrown into sinuous waves, or in the case of leeches flattened into a thin wafer-like form that undulates rapidly in water. Ciliary locomotion occurs in archiannelids.

b Chaetae may act as anchors in polychaetes and oligochaetes, giving purchase on the substrate, in the burrow or on the walls of a tube.

Suckers are temporary points of attachment that enable leeches to remain in one place, to project from a surface whilst awaiting prey, or to allow the looping, stepwise progression that is the typical, slow leech movement. The coelom is largely occluded in

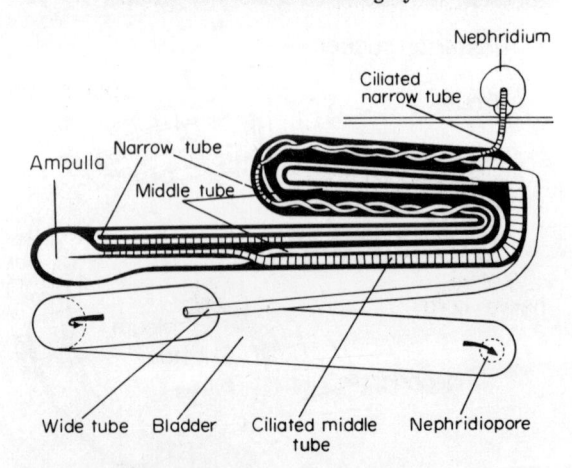

Fig. 17.8 Schematic diagram of arrangement of a nephridium from *Lumbricus* (oligochaete) showing ciliated regions (banded) and relative diameters of the various tubular portions (figure after Graszynski, nomenclature after Ramsay).

this group and movement does not involve large-scale shunting of fluid in the body. The function of the hydrostatic skeleton is taken over by the parenchymatous botryoidal tissue.

c Fast and slow muscles are described for annelids. Inhibitory mechanisms also exist.

Co-ordination

a i The elongate and segmented nature of the annelid body has imposed certain fundamental features upon the nervous system. These include through conduction pathways extending from head to tail, and the repetition of the sensory and motor apparatus within each segment.

The central nervous system has a cerebral ganglion anteriorly, connectives that pass around the pharynx, and a long ventral nerve cord with segmental ganglia. The nerve cord is double and solid. Interneurones pass between the ganglia. Giant fibres (having great diameter relative to other fibres) are found in many species, and act as rapid conduction pathways. The arrangement in the earthworm is shown in Fig. 17.9. In this case the elements of the giant fibres are segmental and abut against their neighbours; electrotonic junctions occur between the lateral giants. In *Myxicola* the giant fibre is greater in diameter than the remainder of the nervous system. The giant fibres seem better developed in burrowing and tubicolous worms. Giant cells (Retzius cells) are found in leech ganglia.

Anteriorly sense organs are usually numerous, especially in predatory polychaetes, in burrowing but active oligochaetes, and in the ectoparasitic leeches. Tubicolous polychaetes may have fewer sense cells, associated with the loss of cephalization. Sense organs are also distributed along the body, and there may be specialized regions such as nuchal organs (chemoreceptors) and eyes, in a number of segments.

Segmental nerves supply the sense organs and the musculature of the body wall. Sense organs include photoreceptors, mechanoreceptors, chemoreceptors, proprioceptors and thermal receptors (in leeches). The fine structure of some of these systems is now known, but function is still difficult to correlate with structure.

ii Fibre patterns and interactions have been studied for many years. In *Hirudo* the number of nerve cell bodies within each segmental ganglion can be estimated and a certain number can be identified. Some 26 are motor cells, whilst 14 are mechano-receptors with known characteristics and centrally-placed cell bodies. These have specific effects upon each other as well as upon interneurones and motor

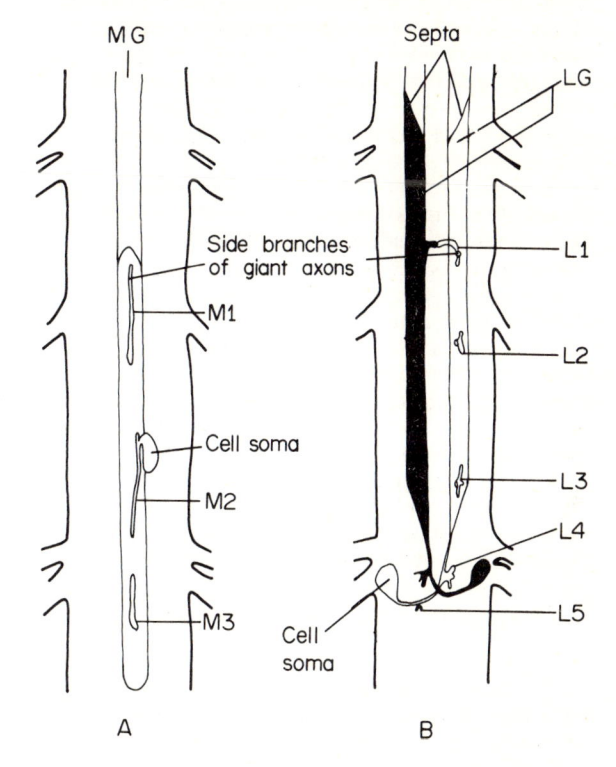

Fig. 17.9 A. Diagram of a median giant (MG) axon in one segment, aligned correctly relative to the structures of the lateral giant (LG) axons. The cell body is displaced somewhat from its midventral position to enable branch M2 to be drawn in place. The septa, which occur near the intersegmental boundaries, mark the terminations of the MG axon. The scale is the same as in B.

B. Diagram of a pair of LG axons in one segment. One cell is outlined, the other solid. The septa are the anterior and posterior boundaries of each cell. One of the five branches of each LG (L1), and the neurite, cross the midline and contact their contralateral homologues. L1, L2, and L3 are the first, second, and third branches of the LG. L4 and L5 are branches of the neurite. Each LG axon is 50 μm. in diameter (A and B after Mulloney).

cells. Some of the sense cells provide input to contiguous segments without benefit of interneurones. *Lumbricus* appears to have central sensory cell bodies also, but there are many peripheral sensory cells. There is a certain amount of intersegmental overlap of mechanosensory information with one segmental nerve involved in receiving stimulation from parts of two contiguous segments. Motor activity is generated by relatively few cells, and these are connected with particular muscles, e.g. two large, fast motorneurones are linked functionally by an electrotonic synapse (so that they show synchrony) and innervate the fast, longitudinal muscle of each segment (*Hirudo*).

The organization of the segmental innervation is concerned with two events, one co-ordinating the movements of the segment and the response to stimulation, and the second transmitting information from that segment to its neighbours and further along the nerve cord.

Through conduction to the extremities is a function of interneurones. These may be giant fibres (or in some cases small diameter fibres as in leeches). These fibres receive direct input from sense organs and provide for rapid response in the musculature. Where there are parallel pathways these may be coupled by electrical connections that are of low resistance. Giant fibres adapt rapidly. Electrical and chemical synapses are known. Dopamine, 5-HT, acetylcholine and γ-amino-butyric acid have all been implicated as transmitters.

b Hormones: some endocrine phenomena are described for annelids in all three classes. The presence of neurosecretory cells, or at least nerve cells that double as secretory cells from time to time, has been demonstrated in all ganglia (cerebral, suboesophageal and nerve cord). The sites of release for the products of these cells have not been described. Only one candidate for a specific release site has been mentioned, the infracerebral gland which lies on the ventral surface of the posterior brain in *Nereis*, where there is close conjunction between the nerve fibres, the coelom and the blood system. In other locations in the nervous system a number of categories of neurosecretory cell have been identified; some are active when the worm is young, others when fully developed, and yet others when maturation is occurring.

The secretory products, no matter how and where they are liberated, have been shown to be concerned with the processes of regeneration, diapause, maturation of the gonads, growth and development, production of the heteronereis (epitoke), colour change and ionic control. A number of types of cell exist and these presumably have different functions.

Respiration

a Gaseous exchange occurs across the general body surface in many annelids but some specialized areas (gills) are found in polychaetes and a few oligochaetes. These are well-vascularized structures in which the blood system comes close to the body surface. In most oligochaetes the skin is well provided with capillaries but gills are found in *Dero* and *Branchiodrilus*, whilst *Alma emini* uses its rear end as a kind of lung. Leeches rely upon the body surface.

Polychaete gills are found amongst the parapodia (*Arenicola, Nephthys*), as thin-walled, branched organs carried anteriorly (terebellids), or as tentacles (*Sabella*) amongst other types. Polynoid polychaetes (*Aphrodite*) pump water in and out of a space between the elytra and the covering felt of chaetal elements.

b Behavioural activity is correlated with respiratory requirements in many polychaetes, particularly burrowers and tube-dwellers. Periodic bursts of irrigation and passage of water through tubes is well known for such genera as *Arenicola, Chaetopterus, Sabella* and *Terebella*.

Most annelids regulate oxygen consumption independently of external O_2 tension, above about 50 mm. Hg pressure. Polychaetes average about 50% uptake of O_2 from water moving across the gills.

Circulation/coelom

a In polychaetes and oligochaetes the coelomic space is large and fluid-filled. It is composed of a series of enclosed pouches that occupy much of the volume of each segment. The whole region is bounded by a peritoneum which derives from the mesodermal layer formed by cell 4d during early growth. During development mesodermal bands appear on each side of the larva, cavities develop, and each pair of cavities (bilateral) eventually form the coelomic space of a segment. These spaces are virtually continuous with one another, separated only by a membrane suspending the gut dorsally and by a thin membranous septum. Each coelomic sac opens to the exterior via coelomoducts, typically a pair in every segment. In some examples the segmental organization of the coelom is modified by loss of septa.

Hirudinea have much reduced coelomic spaces with the cavities being largely invaded by connective tissue and parenchyma (botryoidal tissue). The coelom is restricted to a system of sinuses which may subserve a vascular function and which in some cases becomes contractile to assist in circulation. The coelomic system is closely allied to the blood system, generally enclosing the vessels of the latter, particularly the dorsal and ventral vessels.

Annelids possess a well-developed blood vascular system. Basically, there is a dorsal vessel, a ventral vessel, and connective vessels and capillaries that surround the gut and provide a pathway for blood to move between the main longitudinal channels. Valves are present to prevent backflow. Many organs are well supplied with vessels, especially the skin, gills, nephridia and gut. In polychaetes there is much modification with distension of vessels into sinuses, particularly amongst tubicolous forms, and a few small forms lack a blood system altogether. In leeches there is a similar plan to oligochaetes, with appropriate changes in the regions of the suckers.

b The vessels of annelids are contractile and provide propulsive power for the blood. A great deal of blood movement, however, must be promoted by the contraction of body wall muscles, as shown in Fig. 17.10. Some connective vessels are especially contractile (the lateral hearts of *Lumbricus* and *Arenicola*). Blood moves forward in the dorsal vessel, and backwards in the ventral vessel. The peristaltic wave is probably myogenic in origin.

c Small annelids have colourless blood but amongst large representatives, haemoglobin or the related greenish compound chlorocruorin (in sabellids, serpulids, flabelligerids) may be present. These pigments may be in solution in the plasma (in *Lumbricus*) or in corpuscles (*Travisia, Glycera*). Coelomic fluid contains no pigmented compounds except in species with degenerate vascular systems.

The respiratory pigments have a range of unloading tensions, and variable Bohr effects on changing pH, dependent upon the species considered.

Reproduction

a The gonads of polychaetes usually occur sequentially through a number of segments of the body (Fig. 17.11A). The gonads are not always well-formed organs but may exist as isolated areas on peritoneal and mesentery strands. A variety of methods of release of the gametes is found. The coelomoducts or nephridia may act as vents to the exterior but frequently the body wall ruptures or the gut may erode as in *Platynereis megalops*.

In oligochaetes reproductive organs are located within a few segments only (Fig. 17.11B,C), the animals are hermaphrodite, and gametes are released via coelomoducts. Intromittent organs (penes) are present in some worms (e.g. *Alma, Eutyphoeus*). Leeches are also hermaphrodite (in some groups protandrous) and may possess a penis or other ejaculatory organs (Fig. 17.11D). Sperm may be collected and deposited in spermatophores, and these can be implanted in the skin of the partner during mating.

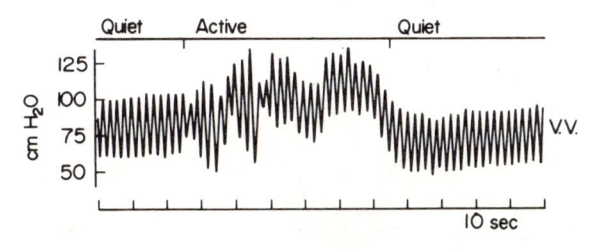

Fig. 17.10 Recording of pressures in the ventral vessel of *Glossoscolex giganteus* (oligochaete) during periods of activity and quiescence. The lateral hearts beat in synchrony and the pressure pulses are smooth. Contractions of body wall muscles double pressures in the vessel (after Johansen & Martin).

b *Sexual reproduction* in polychaetes is usually by external fertilization in the sea, but considerable behavioural and structural modifications occur to ensure maximal efficiency. Many bottom-dwellers bud migratory, swimming, sensory individuals, the 'epitokes', that act as motile gamete carriers for a very brief period, shed the sperm and ova, and then die. Syllids and nereids provide spectacular examples. The short life expectancy of these reproductive fragments is often coupled with a remarkably precise breeding period in which male and female organisms are mature and active at the same time (as short as a few minutes, or perhaps days).

Sexual reproduction in oligochaetes involves copulation, cross-fertilization, storage of sperm, laying of eggs into a cocoon (produced by the clitellum) and the development of a yolky egg that provides nutrient for direct development without a larval stage. Leeches are not dissimilar in reproductive habits, depositing spermatophores which may only release sperm later. This is related to egg-laying that is delayed after copulation ceases (a few days in *Erpobdella*, up to 9 months in *Hirudo*). Both oligochaetes and hirudineans produce a cocoon.

c *Asexual reproduction* by fragmentation occurs, especially among polychaetes, but also in a few aquatic oligochaetes. This takes place by the budding of new individuals (stolons) from a basic stock but there are many variations on the theme (Fig. 17.12).

Cleavage

The embryology and development of annelids is an important aspect of the biology of this phylum.

The pattern of cleavage is spiral and the fate of the cells is determined, from the earliest moments. Cleavage is usually total but unequal since the eggs are yolky. This gives rise to small cells (micromeres) and large yolky cells (macromeres) in the early stages.

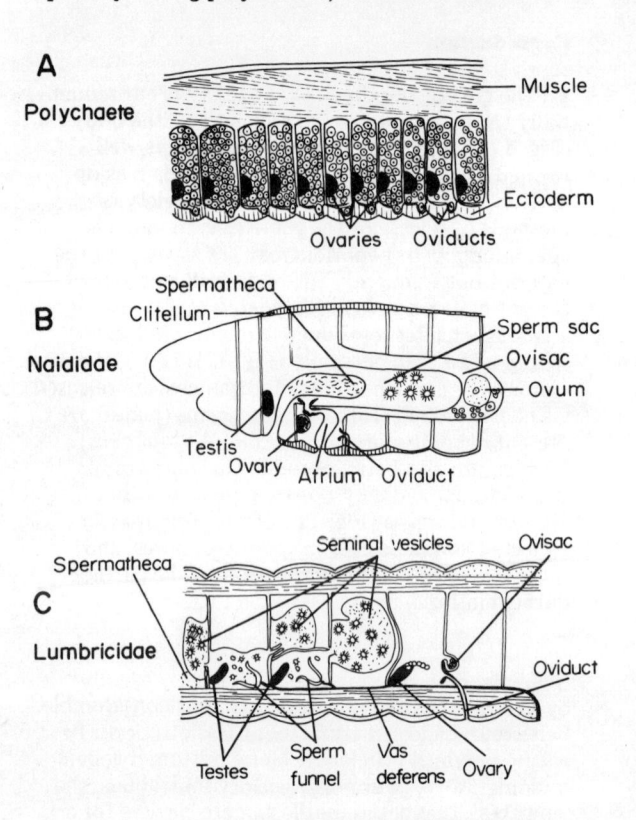

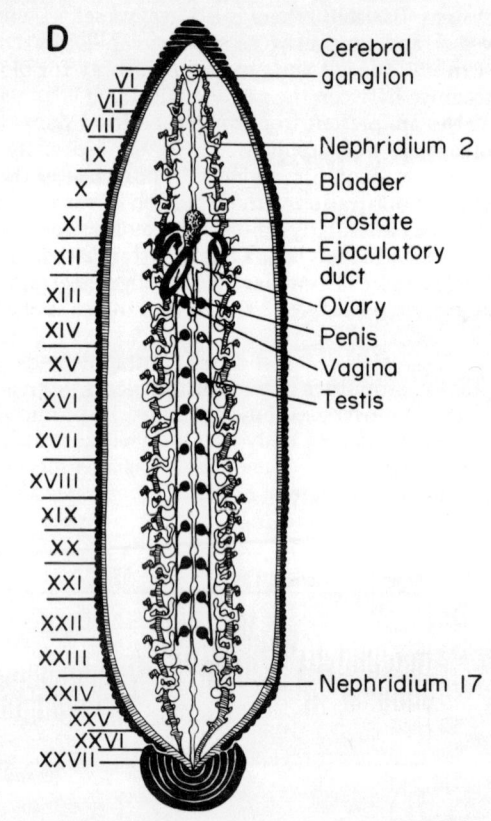

Fig. 17.11 Sexual reproductive organs. A. A polychaete (*Serpula*) with serially repeated gonads. B. An oligochaete (Naididae) with reproductive system massed in 5 segments only. C. An oligochaete (*Lumbricus*) in which there is more spatial separation of gonadal elements and accessory structures (A—C after Borradaile, Eastham, Potts & Saunders). D. Hirudinea: *Haemopis*, showing nephridia and reproductive system (after Mann).

Phylum ANNELIDA

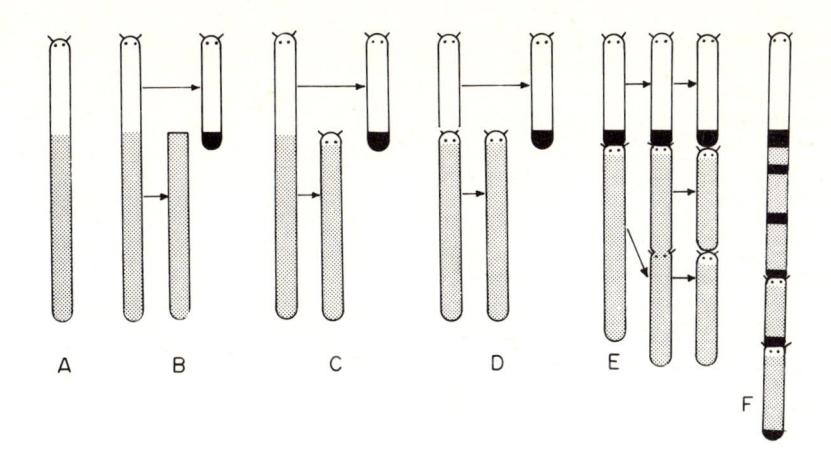

Fig. 17.12 Budding and stolon formation in polychaetes (gamete-bearing part stippled, pygidial regeneration black). A. An epitoke (heteronereid or heterosyllid). B. Gamete-bearing region breaks off (palolo). C. Stolon regenerates head after autotomy. D. Formation of new head and pygidium of stock before freeing stolon. E. Formation of second stolon before liberation of the first. F. Multiple stolons (A—F after Dales).

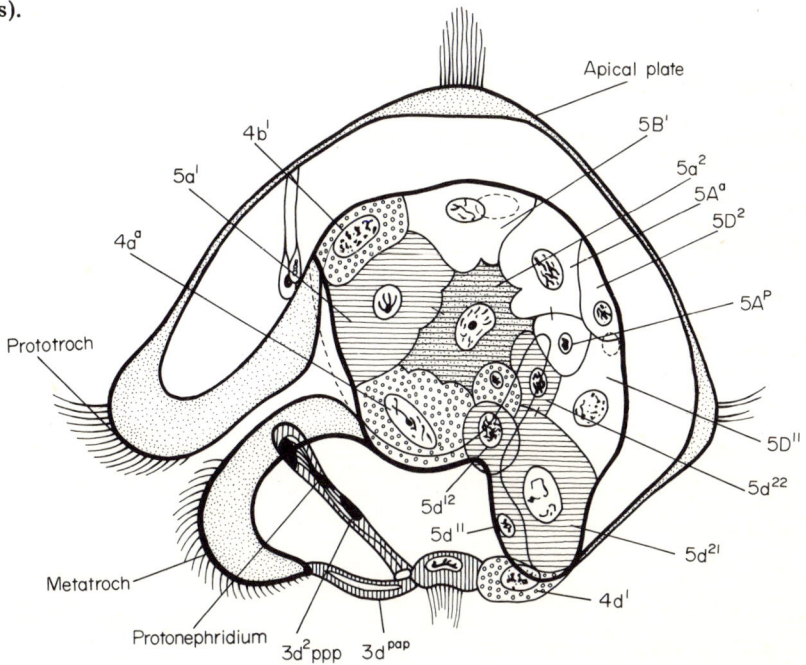

Fig. 17.13 Optical section of young trochophore of *Polygordius* after gastrulation. Numbers, e.g. $5d^{21}$, indicate the derivation of individual cells (see text) (after McBride).

Phylum ANNELIDA   85

By convention the development and fate of the individual cells of the embryo may be traced and they may be identified and labelled in a rational sequence (Fig. 17.13). Very briefly this may be stated as follows:

the *1st* division gives rise to 2 cells;

the *2nd* division gives rise to 4 cells, labelled A,B,C,D;

the *3rd* division occurs horizontally so that 4 upper pole micromeres separate from 4 lower, larger macromeres (vegetal pole). The micromeres are 1a, 1b, 1c, 1d and the macromeres are 1A, 1B, 1C, 1D;

the *4th* division: if the first division plane was clockwise the second is counterclockwise and this division gives a further quartet of micromeres and macromeres which are 2a, 2b, 2c, 2d and 2A, 2B, 2C, 2D respectively.

This process continues through a series of 3A, 4A, etc. After 4A the macromeres are smaller than the micromeres.

During these divisions the earliest micromeres themselves divide and the cells thus formed are labelled.

1a gives rise to $1a^1$, $1a^2$. When $1a^2$ divides $1a^{21}$ and $1a^{22}$ appear. $1a^{22}$ gives rise to $1a^{221}$ and $1a^{222}$. Similar nomenclatures are given to all cell lineages.

Each cell gives rise to a structure or group of structures that is predetermined. Cell 4d, for example, gives rise to presumptive mesodermal structures (Fig. 17.13).

References

Dales R.P. 1963. *Annelids*. Hutchinson University Library, London.
Edwards C.A. & Lofty J.R. 1972. *Biology of Earthworms*. Chapman & Hall, London.
Mann K.H. 1962. *Leeches*. Pergamon Press, Oxford.

18 Phylum ECHIURIDA

Small to medium-size worms up to 10—15 cm in length. About 80 species known.

Characteristics

1 All are marine, most sub-littoral but a few inhabit deeper water.
2 The larva is a trochophore.
3 The adults are unsegmented.
4 Possess a large, non-retractable proboscis.
5 Have one pair of ventral chaetae.

Larval form

The larva is a typical trochophore resembling that of polychaetes.

Metamorphosis

Growth of the trochophore to the adult condition is also like that of polychaetes. There is a pre-oral lobe with pre- and post-oral ciliary bands (Fig. 18.1). The posterior region elongates and the mesoderm blocks are segmentally arranged (up to 15). The nervous system briefly demonstrates metamerism. Chaetae develop. The signs of segmentation disappear with advancing growth and the pre-oral lobe eventually forms the adult proboscis.

Adult body form e.g. *Bonellia, Ikeda, Urechis.*

The adult is unsegmented, and possesses one pair of ventral chaetae just posterior to the prominent proboscis. Other chaetae may be found at the posterior extremity. The proboscis is non-retractable, variable in length and very mobile. It may bifurcate (*Bonellia*) (Fig. 18.2A), be extremely long relative to body size (*Ikeda*), or be relatively small (*Echiurus*) (Fig. 18.2B). It is usually guttered on the ventral side. The body is rounded and smooth or papillate, very well muscularized, and contains a spacious coelom (Fig. 18.3A, B). The skeleton is hydrostatic.

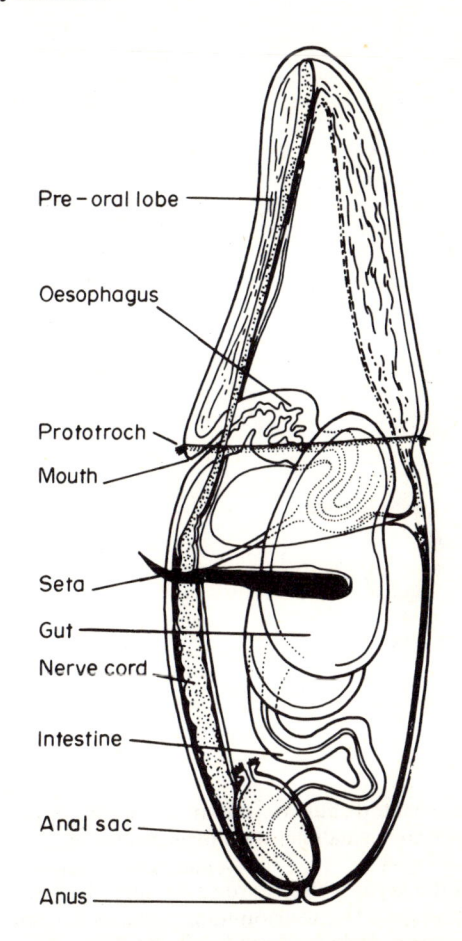

Fig. 18.1 Metamorphosing larva of *Echiurus* indicating development of proboscis as a pre-oral lobe, and the position of chaetae post-orally. Anal sacs are also beginning to appear (after Baltzer).

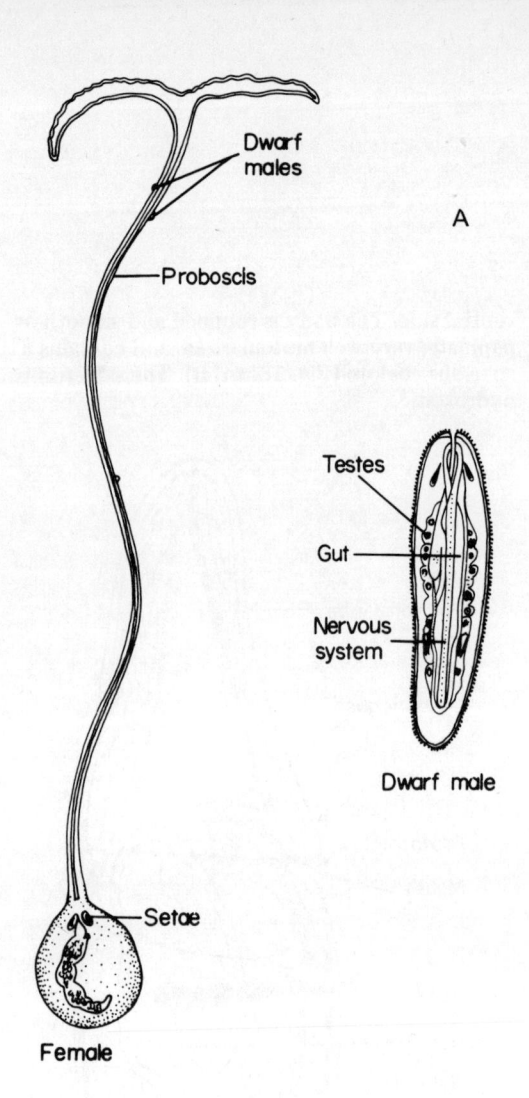

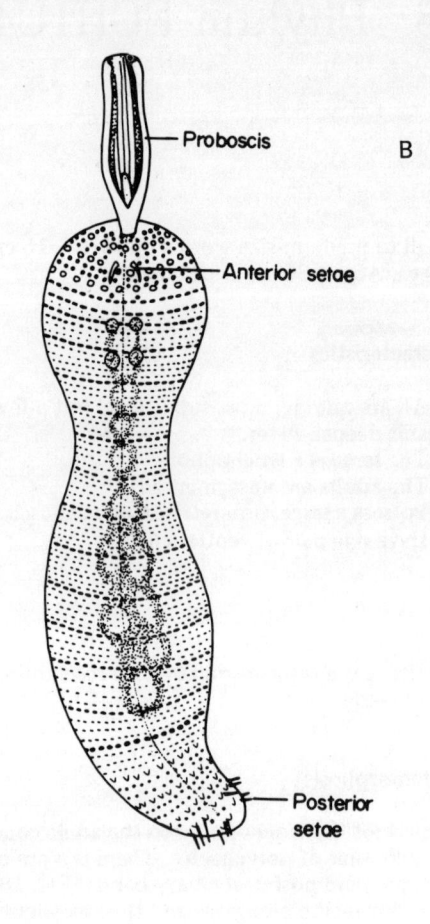

Fig. 18.2 A. *Bonellia viridis* (after Baltzer). B. *Echiurus* (after Greef).

Feeding

a Most echiurids burrow in mud and sand, or are incarcerated in small natural openings amongst rocks and shells. One, *Thalassema*, inhabits dead sand dollar shells which it grows too large to leave. All obtain food from surrounding surfaces by using the proboscis. Detritus from the immediate surroundings is gathered by a mucous secretion that either forms nets or funnels through which water is drawn (*Urechis*) (Fig. 18.4), or to which particles adhere as the proboscis combs the substrate (*Bonellia*, *Echiurus*).

b The proboscis is the feeding organ, picking up material and passing it by ciliary action to the mouth. The mouth lies at the base of the proboscis, the anus is terminal. The gut takes a tortuous, coiled path, especially the intestine and hind gut (Fig. 18.3B). Alongside the intestine runs a small tube which opens at both ends into the ciliated intestinal groove.

88 Phylum ECHIURIDA

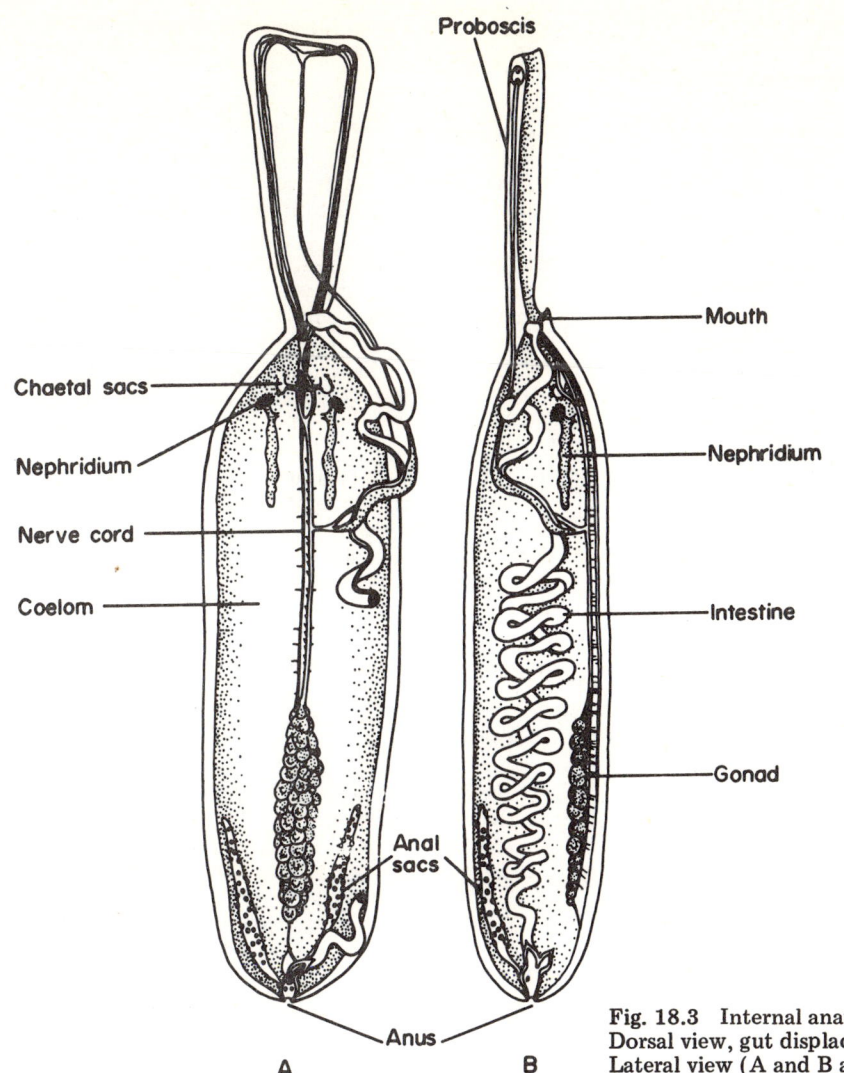

Fig. 18.3 Internal anatomy of an echiurid. A. Dorsal view, gut displaced and intestine removed. B. Lateral view (A and B after Delage and Herouard).

Osmoregulation/excretion

a   The excretory metanephridia are large sacs, occupying considerable space within the body. In *Ikeda* hundreds of pairs exist whilst in other species there are few pairs (one in *Bonellia*, two in *Echiurus*). Sexual dimorphism is exhibited in this feature, *Thalassema* males having more nephridia than females. In *Bonellia* females the dwarf males may attach to the nephridia within the coelom. Contractile, elongate anal sacs are equipped with many ciliated ducts opening to the coelom. These void contents via the anus whilst the nephridia open anteriorly.

b   Excretory and regulatory physiology is not known.

Phylum ECHIURIDA

Movement

a Adult echiurids show no movements outside the permanent residence. The proboscis is often muscular and capable of great extension and shortening, being used in exploration of the substrate surface. In one genus, *Epithetosoma*, the coelom pushes up into the proboscis, providing a hydrostatic skeleton. Within the burrow most species show minor adjustments of position but in *Urechis* pumping of water occurs by rhythmic peristaltic motion of the body, moving about 29 litres water/day. Migrations along the burrow also occur. There is a startle response.

b The body wall is well endowed with powerful muscles.

c *Urechis* governs internal hydrostatic pressures by adjusting the volume of water in the hind gut, balancing pressures generated by muscular contractions against inhalent movements of water into the hind gut. The internal pressure decreases when the anus opens following arrival of a peristaltic pulse from the proboscis. Forced exhalent currents relieve the pressures within the gut, and also clear the burrow of its contents.

Co-ordination

a *i* The nervous system consists of a single, solid ventral nerve cord that bifurcates rostrally to form two circumpharyngeal connectives that unite in the proboscis. From the ventral cord arise many hundreds of segmentally-arranged peripheral nerves, sometimes opposite, sometimes alternate and sometimes irregular in position. The nerve cord may be sinuous. Ganglia are lacking. There is a multicellular giant fibre (*Urechis*). Sense organs are sparse.

ii The giant fibre conducts at about 1.5 m/sec in both directions. The remainder of the nerve trunk is much slower-conducting. Pacemakers responsible for anal pumping, and for peristalsis, are located in the nerve cord. These activities are primarily initiated in the proboscis.

b Hormones are unknown.

Respiration

a The hind gut may be active in respiration, conducting regular intake and output of water in *Urechis*. Otherwise the body wall is the main site of exchange.

b Increased frequency of peristaltic waves starts when oxygenated sea water bathes the animal (Fig. 18.5).

Circulation/coelom

a There is a closed blood vascular system, except in *Urechis*. The plan is similar to that of annelids. The coelom occupies a large volume.

b Blood flow is promoted by small contractions of circum-intestinal vessels and by body wall movements. Coelomic fluid is also involved in circulation.

c The blood is colourless, but contains amoebocytes. Some cells in the coelom contain haemoglobin and may assist in oxygen transport.

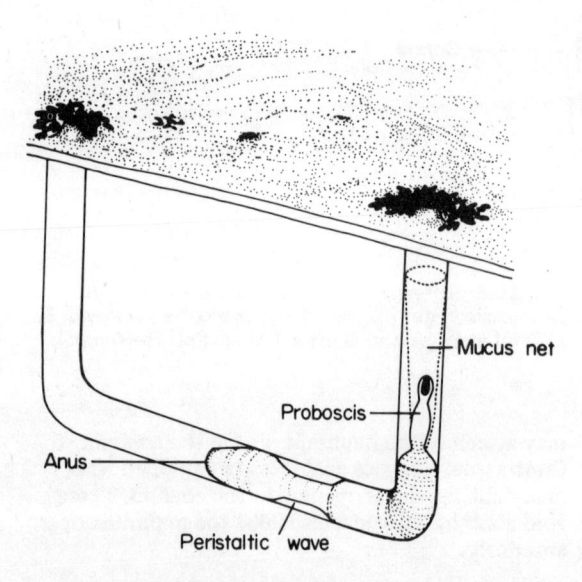

Fig. 18.4 The position of *Urechis* in its burrow, showing peristaltic waves along the body, and the anterior mucus net (after Lawry).

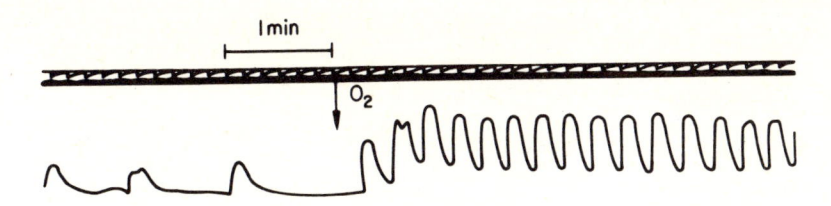

Fig. 18.5 The peristaltic movement of the body, and various organ components, is stimulated by the presence of oxygen in the environmental water. In this case the proboscis response is shown: oxygen added at arrow (from Lawry).

Reproduction

a The sexes are separate. The gonads are located on the peritoneum of the ventral mesentery in the trunk. Gametes are released into the coelom where they mature before escaping via the nephridia.

b Fertilization is external except in *Bonellia*. In this genus, as mentioned above, the dwarf males are typically attached to the nephridia, and the eggs are fertilized within those organs. Extreme sexual dimorphism is shown; the minute, ciliated males have no proboscis or blood system, and possess only a vestigial gut. Cleavage is spiral.

Affinities

Although classified as a separate phylum because of numerous specializations and pecularities, echiurids are related to annelids and sipunculids. The major points of resemblance are shown in Table 18.1.

Table 18.1 Echiurid-Annelid-Sipunculid affinities (from Clark).

	Echiurida	*Annelida*	*Sipunculida*
1	Early development and structure of the early trochophore similar in all groups		
2	Possible metameric organization at the earliest mesodermal appearance	Metameric	No signs of metamerism at any stage
3	Adult unsegmented	Adult segmented	Adult unsegmented
4	Ventral nerve cord arises as a double row of cells (later united)	Not fused	Nerve cord arises as a single strand
5	Chaetae secreted by a single chaetoblast cell	Single cell	Chaetae formed by numerous cells
6	Trochophore possesses protonephridia replaced by metanephridia; anal sacs in echiurids have parallels in gut-opening nephridia of oligochaetes	Protonephridia and metanephridia	No protonephridia No anal sacs
7	Blood vascular system nearly identical in echiurids and annelids		
8	Gut has a ciliated gutter (ventral)	Typhlosole	Gut has ciliated gutter (dorsal)

References

Clark R.B. 1969. Systematics & Physiology: Annelida, Echiura, Sipuncula. *Chemical Zoology*, Vol. 4 (Eds. Florkin & Mason). Academic Press, New York and London.

Marshall A.J. & Williams W.D. 1972. *Textbook of Zoology, Invertebrates* (7th Edition of Parker & Haswell, Vol. I). Macmillan, London.

19 Phylum SIPUNCULIDA

About 250 species of sipunculids are extant. They are small (0.2 mm) to medium (50 cm)-sized animals.

Characteristics

1 The group is entirely marine.
2 The larva is a trochophore.
3 The adults are unsegmented and elongate, worm-like animals.
4 They are coelomate.
5 There is no prostomium in the adult.
6 The anterior portion of the body forms an introvert that may be retracted into the trunk region. The terminal mouth is surrounded by tentacles.
7 The gut is U-shaped, opening at the anus which is dorsally and anteriorly placed.
8 There are one or two metanephridia.

Larval form

The larva is a trochophore (Fig. 19.1A). This is similar to that of annelids. The egg membrane (vitelline) encloses the larva as a cuticle until metamorphosis begins. Swimming is vigorous and the animal is photopositive.

Metamorphosis

In *Phascolosoma* the trochophore metamorphoses at the end of the second day, becoming elongate and showing muscular activity. It becomes bottom-living (Fig. 19.1B).

Adult body form e.g. *Sipunculus, Phascolion*.

There are no classes erected for the Sipunculida and the anatomy of *Phascolosoma* or *Dendrostomum* may be taken as typical. The body form is shown in Fig. 19.2.

The introvert is the anterior portion of the body which is retracted, under the influence of large retractor muscles, into the trunk region (Fig. 19.3). When it is extended the mouth lies centrally and terminally and is surrounded by tentacles (the frilled organ).

The body wall is cuticularized; no chaetae or other bristles are found but in *Phascolion* a holdfast is present and in some other examples spines may be found. The tentacles around the mouth have internal cavities, which are continuous with sacs around the oesophagus that act as compensation reservoirs when the tentacles retract. The skeleton is hydrostatic and the coelom is a continuous sac internally.

Feeding

a Little is known about feeding. Ciliated grooves on the tentacles may be involved in a ciliary-mucous type of food collection, but some sipunculids ingest great quantities of mud and sand, and food may be simply obtained during the processes of burrowing.

Diatoms, protozoa and perhaps other small invertebrates seem the most likely food.

b The gut is recurved with mouth and anus close to one another. The intestine is frequently spirally arranged (Fig. 19.3).

Osmoregulation/excretion

a Metanephridia are located within the ventral anterior trunk (see Fig. 19.3). Normally there is a pair, but in *Phascolion* there is only one. The nephrostome is a simple opening of the tubed organ attached to the body wall. In many species the rest of the nephridium hangs freely in the coelom, but in some it is attached by a mesentery.

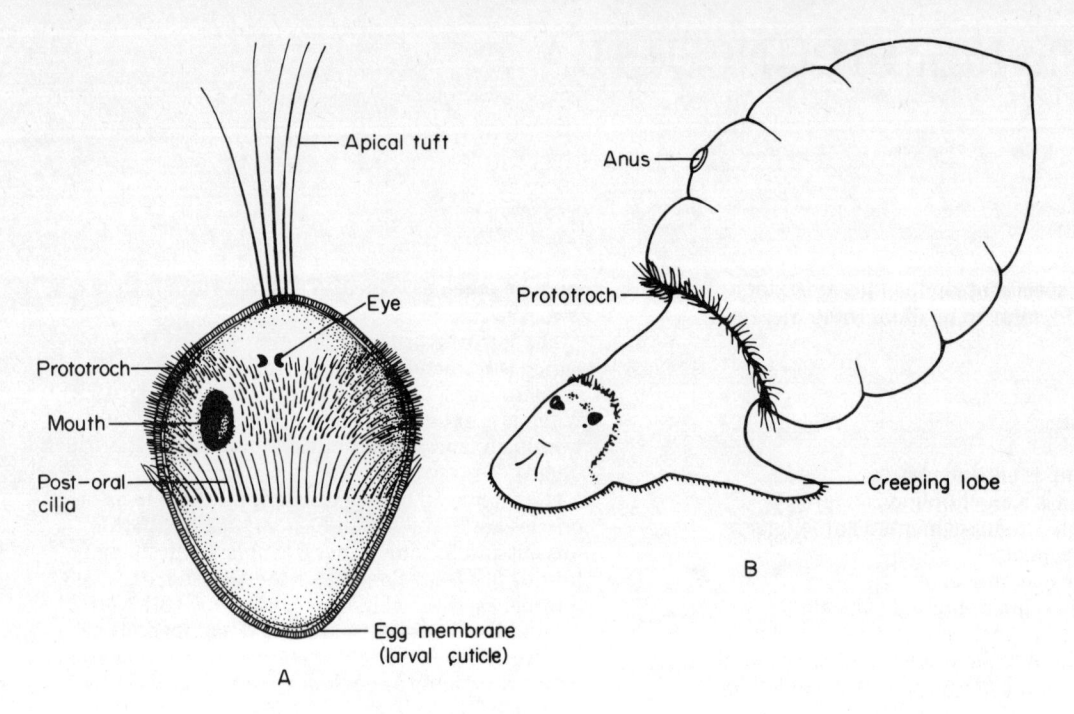

Fig. 19.1 A. Trochophore of *Phascolosoma* (after Gerould). B. Planktonic larva (pelagosphaera) of *Phascolosoma* sp. at the stage when a creeping foot is present (after Jägersten).

b Sipunculids are osmotic conformers. The coelomic fluid has approximately the same osmotic pressure as sea water; the body weight increases when placed in dilute sea water, and decreases in more concentrated sea water. Some species show ion control through the gut and nephridia. The osmoregulatory function of the nephridia is indicated by the fact that weight increase is greater and subsequent readjustment poorer when the nephridia have been removed. The removal of formed elements from the coelom may also play a part in excretion.

Movement

a Sipunculids are sedentary animals, with a preference for burrows and other narrow tubes. They are found in mud, sand, gravel, shell, coral, sponges and other habitats. In all cases burrowing is the main movement exhibited, as a function of the extroverted tentacles and the hydraulic ram effect

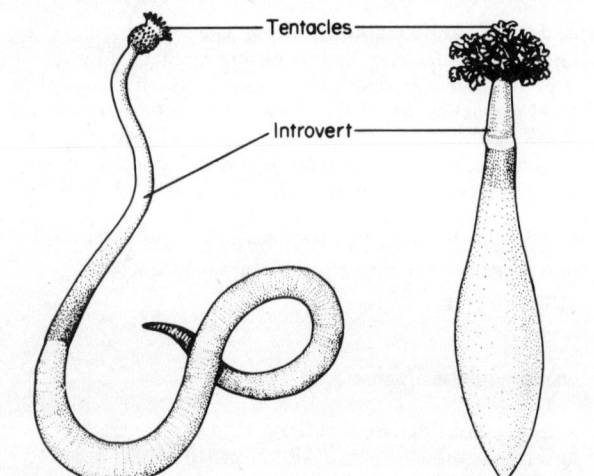

Fig. 19.2 A. *Phascolosoma vulgare* (after Selenka). B. *Dendrostomum pyroides* (after Fisher).

Phylum SIPUNCULIDA

of the body. When established in a burrow the tentacles are extended from the aperture, withdrawing rapidly upon tactile stimulation. Some crawling has been reported, as also have swimming, and righting movements. Positive thigmotropism is the predominant behaviour pattern.

b Movement is brought about by the hydrostatic skeleton working against the muscular body wall, and the major retractor muscles. These muscles are very long and strong and bring about the inversion of the tentacles.

c The retractor muscles show phasic (fast) and tonic (slow) contractions, but it is uncertain whether this is due to 2 types of muscle fibre.

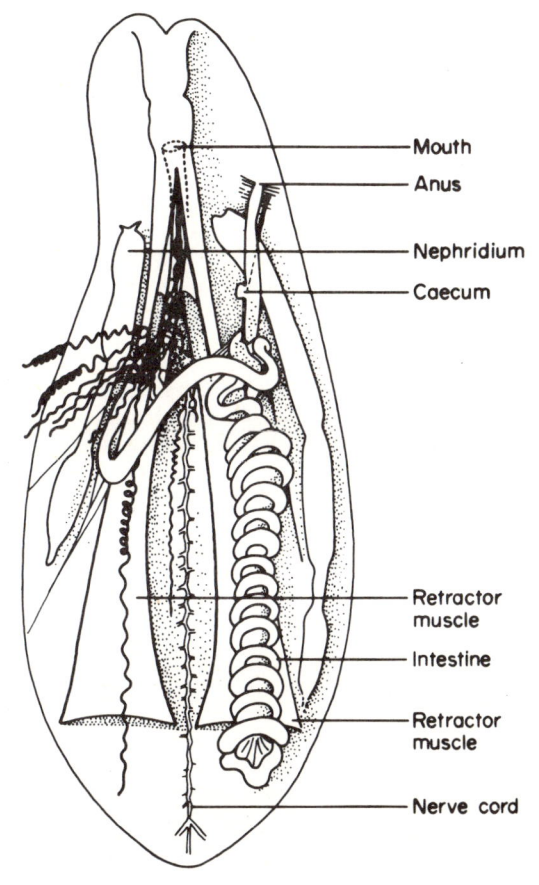

Fig. 19.3 Internal anatomy of *Dendrostomum pyroides* (after Fisher).

Co-ordination

a *i* The brain lies anteriorly above the oesophagus with circumoesophageal connectives joining ventrally to the single ventral nerve cord. This nerve cord is not segmentally organized and there are no ganglia. Sipunculids are well endowed with sense organs: there are photoreceptors in the brain (fine structure like those of polychaetes), nuchal organs (presumptive chemoreceptors), and many sensory papillae, especially on the introvert.

ii Virtually nothing is known of the physiology of the nervous system, but the brain is believed to be inhibitory on the remainder.

b Hormones are unknown.

Respiration

a There are no special respiratory organs, but the expanded tentacles must be of some significance since they bring the internal coelomic fluid close to the external watery environment.

b Some data are available as to oxygen consumption.

Circulation/coelom

a There is no blood vascular system, but the coelom is massive and continuous throughout the body. The coelomic fluid circulates freely in all regions, including the tentacles.

b Propulsion is due to the action of ciliated cells lining the coelomic cavity, and the muscles of the gut and body wall.

c Although not a haemocoel homologous with others, the coelom in sipunculids is the major circulatory system. There are many cellular components in the coelomic fluid, including corpuscles containing the red respiratory pigment, haemerythrin, that acts as an oxygen reservoir. Also in the coelomic fluid are cell groups known as urns. These are initially fixed to the peritoneum (Fig. 19.4), but later become free and concentrate excretory material within.

Phylum SIPUNCULIDA

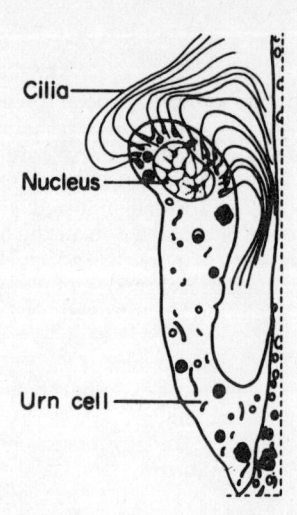

Fig. 19.4 A fixed urn attached to the peritoneal wall, showing extensive ciliation (after Volkonsky).

Reproduction

a The sexes are separate, with the gonads located in coelomic epithelium on the retractor muscles. The gametes are released into the coelom, take several months to become mature, and are shed via the nephridia.

b Fertilization is external, spawning being in the summer, the males leading. Development is by spiral cleavage and the resultant larva is a trochophore.

Relationships

Argument about the systematic position of sipunculids has been long-lasting. Recent opinion indicates that although warranting phyletic status, they are related to the annelids. The separation of the group is due to the non-metameric grade of organization but similarities to annelids exist (Table 19.1).

Table 19.1 Contrast and comparison of sipunculids with annelids (from Clark).

Dissimilarities to annelids	Similarities to annelids
1 Non-metameric	1 Prototroch arrangement
2 Position of the apical cross in the embryo (more like molluscs)	2 Egg membrane becomes larval cuticle
3 Non-segmented nerve cord which is single	3 Similar photoreceptors
4 Larval bristles are not structural	4 Similar nuchal organs
	5 Corpora pedunculata in the brain
	6 Origin of the nervous system
	7 Integument sense organs
	8 Position of embryonic bristles

References

Clark R.B. 1969. Systematics & Physiology: Annelida, Echiura, Sipuncula. *Chemical Zoology*, Vol. 4 (Eds. Florkin & Mason). Academic Press, New York and London.

Hyman L.H. 1959. *The Invertebrates*, Vol. 5. McGraw-Hill, New York.

20 Phylum POGONOPHORA

About 100 species so far discovered. Up to several cm in length.

Characteristics

1 Marine animals, free-living, generally in soft sediments.
2 Sedentary, living in a tube secreted by the animal.
3 Bilaterally symmetrical.
4 Adult polymeric (i.e. with a number of segments); generally recognized to be in three portions: forepart, trunk and opisthosoma.
5 Tentaculate.
6 Coelomate.
7 No digestive system or alimentary canal.
8 Closed blood vascular system.
9 Nerve system primitive with a median cord.
10 Separate sexes.
11 Development total, unequal and bilateral.
12 No pelagic larva.

Larval form

Early embryonic development gives rise to a juvenile. There may be some species with free-living larvae.

Metamorphosis

Development incompletely known.

Adult body form e.g. *Oligobrachia*, *Siboglinum*.

The vermiform body shape is maintained by a hydrostatic skeleton in which both the coelom and blood vascular system play a considerable part.
  The organization of the body is shown in Fig. 20.1.

The recent discovery of a small segmented hind part has confused the terminology applied to various regions but three major areas are commonly recognized (see below).

$$\left.\begin{array}{r}\text{Forepart} \\ \text{Trunk}\end{array}\right\} = \left\{\begin{array}{l}\text{Protosome} \\ \text{Mesosome} \\ \text{Metasome}\end{array}\right.$$
Opisthosoma = Posterior segmented region

There is also confusion at present in the literature as to the correct orientation of the body. This largely depends upon the position of the nerve cord. This may be dorsally or ventrally placed depending upon whether a deuterostome or protostome relationship is considered. The position of the brain is apical and hence unhelpful in deciding a dorsal or ventral position. The authorities are not yet in agreement.
  Tentacles are prominent in the structure of Pogonophora. They may be arranged in complex patterns (Fig. 20.2) which reach a zenith in *Lamellibrachia barhami*. This species has 2,000 tentacles arranged to form about 25 concentric lamellae around and attached to a double lophophore.
  There are two recognized orders.
Order *Athecanephria*: anterior coelom sac-shaped; spermatophores cylindrical.
Order *Thecanephria*: anterior coelom horseshoe- or corkscrew-shaped; spermatophores flat.
  Some authors place these two orders in the Class Frenulata (bearing a bridle) whilst *Lamellibrachia* alone constitutes the Afrenulata (without a bridle).
Skeleton. Pogonophora are tube-builders (Fig. 20.3). The organic part of the tube is chitinous and is identical with diatom β-chitin. This is secreted by multicellular pyriform glands of the forepart and trunk, and by unicellular glands of the forepart. The tube is not usually branched. The walls are known to be permeable, at least to water, sodium chloride, sucrose and phenylalanine.

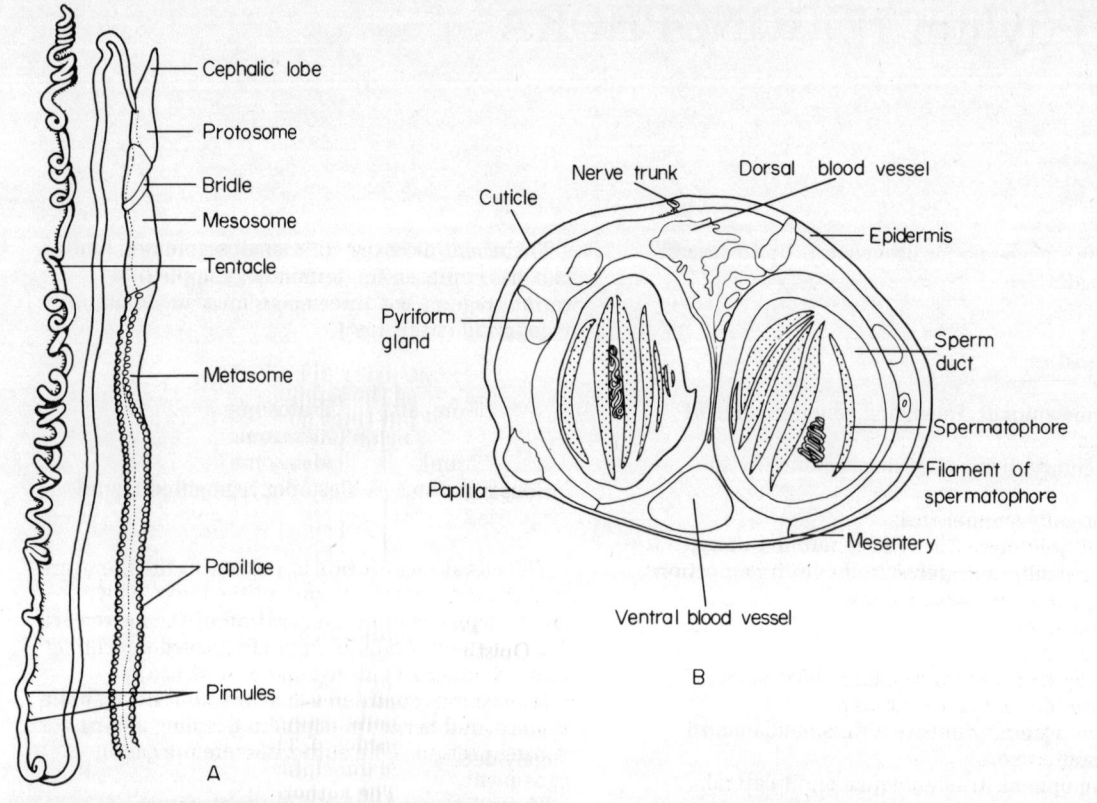

Fig. 20.1 A. Front portion of *Siboglinum caulleryi* (after Ivanov). B. Transverse section of male *Lamellisabella zachsi* in metasomal area. Note absence of a gut (after Ivanov).

Feeding

a The method of feeding is enigmatic. There is no alimentary canal present at any time in the life of the animal.

b No extracorporeal enzymes have been demonstrated so digestion outside the body does not seem to take place.
  Amino acids, glucose and fatty acids are all taken up from the external environment and concentrated in the blood. Protein and ferritin have also been shown to be incorporated from outside and a process of pinocytosis may account for this (although the entire body is covered by a continuous and stout cuticle). Nutrition seems wholly dependent therefore upon the uptake of molecular particles.

Osmoregulation/excretion

Nothing is known about excretion, and there appear to be no special structures.

Movement

Slight movements of the animal up and down the length of the tube are accomplished with the aid of setae present in various regions of the body. In some species there may be burrowing in the substrate.

Co-ordination

a *i* A concentration of nervous tissue is found anteriorly though the precise orientation demands

Phylum POGONOPHORA

further attention. From this area arises a median cord (ventral or dorsal in position according to various authors). The tentacular nervous system consists usually of one main trunk though in some species there are additional small branches. This main nerve trunk contains 4 or 5 large axons and 500–600 small axons. No supporting glial cells have been described.

ii No information is available on the physiology of the nervous system.

b Hormones are not yet known.

Respiration

No information available apart from data on rates of uptake.

Circulation/coelom

a The circulatory system is closed; dorsal and ventral vessels are joined by tentacular and commissural vessels. A 'heart' is developed at the base of the tentacles, associated in some species with a pericardium.

The coelomic space appears as a single anterior

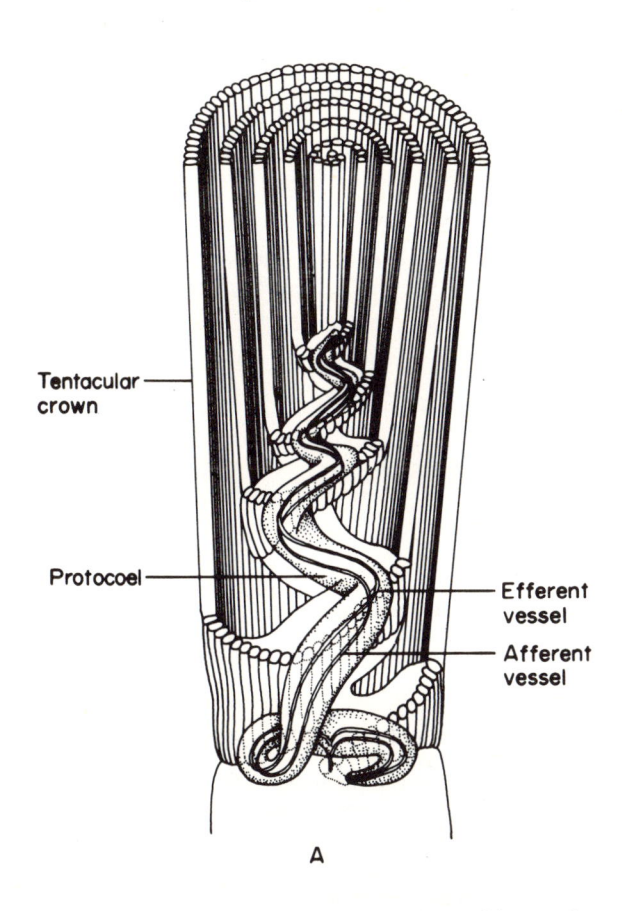

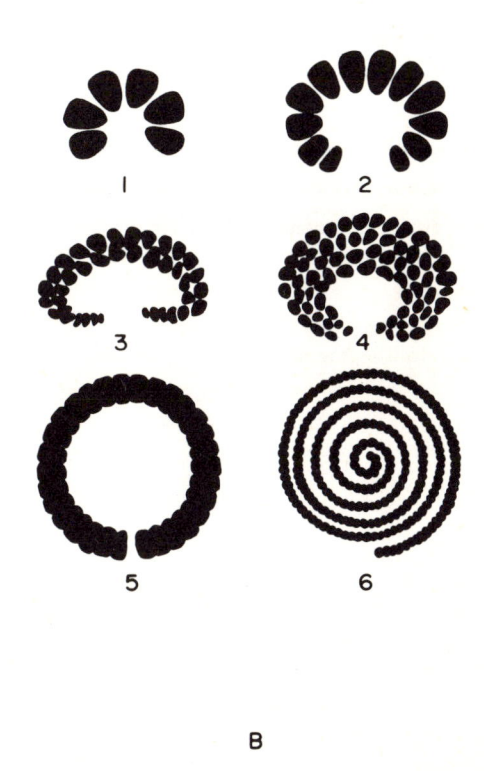

Fig. 20.2 A. Arrangement of tentacles and internal protocoel in *Spirobrachia grandis* (ventral tentacles removed). B. Diagram of arrangement of bases of tentacles in 1. *Oligobrachia dogieli*, 2. *Birsteinia vitjasi*, 3. *Polybrachia annulata*, 4. *P. barbata*, 5. *Lamellisabella zachsi*, 6. *Spirobrachia grandis* (after Ivanov).

Phylum **POGONOPHORA**

Table 20.1 (from Southward). Features demonstrated by Pogonophora, and their similarities to other groups.

Deuterostome features	Annelid features	Remarks
Brain and nerve stem on dorsal side	Nerve stem on ventral side	A ventral brain is not an annelidan feature
Unpaired nerve stem		A few annelids have an unpaired nerve stem
Simple intra-epidermal nervous system		Nerve stem is intra-epidermal in archiannelids, some oligochaetes and a few polychaetes
'Heart' in ventral blood vessel	'Heart' in dorsal blood vessel	
Pericardium present		Pericardium not found in all pogonophores; may be specialization without evolutionary significance
Unpaired anterior coelom		
Paired posterior coeloms	Paired posterior coeloms	
Enterocoelic development of coelom (Ivanov)	Schizocoelic development of coelom (Nørrevang)	
Radial cleavage?		Needs confirmation
Primary division of coelom into 3 segments		
	Later division of coelom into several segments	
	Posterior growth zone	
	Presence of setae	Apparently different mode of formation from annelids
	Setae arranged segmentally	
	Epidermal cuticle consisting of microvilli and layers of fibres	External layers of microvilli are quite common in invertebrates, but layers of fibres have only been found in annelids.
	Chitin of annelid/mollusc type in tube	Annelid tubes do not contain chitin, though setae do

Phylum POGONOPHORA

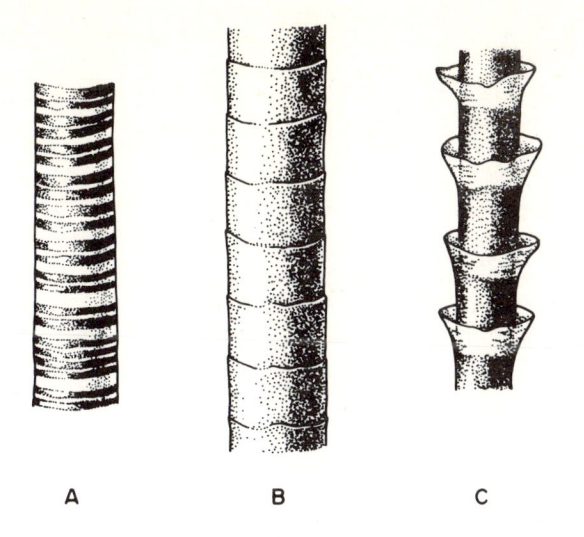

Fig. 20.3 Tubes of A. *Siboglinum*. B. *Lamellisabella*. C. *Polybrachia* (all after Ivanov).

pouch in development that communicates at first with paired, lateral pouches; these then pinch off and later still pinch off a further pair at the rear. This leads to the postulation of a tri-coelomate organization that provides support for a deuterostome relationship. Recent work, however, suggests that even more segmentation occurs, especially towards the rear and hence the animals are polymeric.

b Blood flow is slow in the circulatory system.

Reproduction

a The sexes are separate. The gonads lie in the trunk region. The ovaries lie anteriorly, and the testes posteriorly in the trunk. The ova are large, yolky and usually spherical. The spermatozoa are elongate and bound together in a spermatophore for release.

b Reproduction may be

i sexual with fertilized eggs retained in the tube of the female;

ii asexual (transverse fission occurs in the species *Sclerolinum brattstromi*).

Relationships

Table 20.1 indicates the major similiarities and differences between the pogonophores, annelids and deuterostomes. On present evidence it seems that annelids are perhaps the closest relatives, but further information on the orientation of the pogonophoran body would be of value.

Distribution

Taken from all major oceans (deep water) and some coastal areas, pogonophores are mainly found on continental slopes, in deep trenches and basins and around oceanic islands. This may represent a false picture, however, only indicating where expeditions have so far been successful. The distribution may be worldwide.

References

Ivanov A.V. 1963. *Pogonophora*. Academic Press, New York and London.
Southward E.C. 1971. *Oceanogr. Mar. Biol. Ann. Rev.* 9, 193—220.

21 Phylum ONYCHOPHORA

Length up to 15 cm. About 65 species described.

Characteristics

1 All are terrestrial in damp places, in tropical or south temperate areas.
2 A thin, flexible cuticle is present.
3 Body wall has circular and longitudinal muscle layers.
4 There are numerous stumpy, unjointed, clawed, paired appendages.
5 Coelomoducts are lined with cilia.
6 Development is direct, some species brooding the young.

Larval form

There is no larval stage (see section on reproduction).

Metamorphosis

Nil.

Adult body form e.g. *Peripatus*, *Peripatopsis*.

Onychophorans exhibit some features typical of annelids, and others which are those of arthropods. The external surface is covered by a thin flexible cuticle that is unlike the jointed exoskeleton of arthropods. It is chitinous and not segmented. The cuticle is warty and bumpy. Some 14 to 43 pairs of short limbs (legs) are arranged ventro-laterally along the body. Each bears two claws. There are antennae anteriorly. The body wall is constructed in annelid style with epidermis, dermis, circular, oblique and longitudinal muscles in layers. The mouth is terminal, flanked by glandular oral papillae, and armed with a three-piece tooth apparatus.

Reproductive organs are associated with crural glands that open near the nephridiopores (Fig. 21.1). The skeleton is hydrostatic.

Feeding

a Onychophora are omnivores. They feed on wood fibres, insect excreta, small molluscs, snails and termites. They may even be cannibalistic.

b The mouth region is well-muscularized, ringed by small papillae, and armed with a pair of jaws in which the mandibles are composed of curved chitinous plates. The pharynx is thick-walled and cuticularized, leading to a narrow oesophagus. Salivary glands, secreting protease, are large and their secretions enter the buccal cavity. These glands are specially modified, and the first in a series of excretory organs. The intestine is long, opening via a short thin-walled rectum at a posterior terminal anus.

Osmoregulation/excretion

a Nephridia occur in pairs along the body. Each organ is formed by an internal blind sac (coelomic), a tubular canal and a ciliated duct opening via a bladder ventrally on the leg bases.

b There is no information on the function of the nephridia but peripatoids select high humidity regions, arguing for a high water loss from the body (see section on respiration).

Movement

a Locomotion is by walking on the stumpy legs.

b The legs lift the body clear of the ground. Relative positions are changed by the extension and

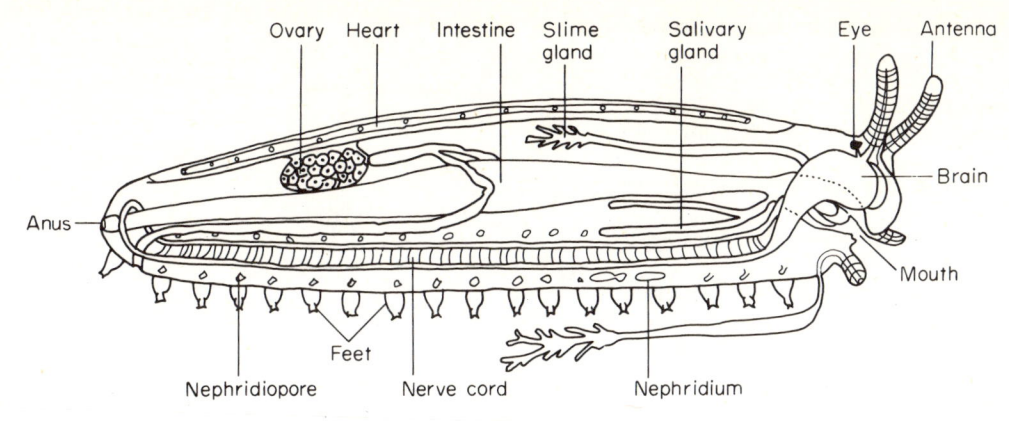

Fig. 21.1 Diagrammatic representation of *Peripatus* to show internal organs (after Cuénot).

contraction of the longitudinal muscles of the body, but leg muscles are concerned with actual positioning of the appendages. Speeds of movement are variable, slow progression being a function of a short body, fast progression that of a lengthened body.

c No details are available on muscle physiology.

Co-ordination

a *i* The nervous system is non-ganglionate. Anteriorly there is a major nervous mass, the brain, lying in the head. Extending caudally are two longitudinal nerve cords, connected by commissural nerves. Nerve cells are distributed along the length of the cords without ganglia being formed, although slight swellings are noted at the base of each pair of legs. Sense organs include large eyes (at the base of each antenna), putative chemoreceptors and mechanoreceptors on the antennae, and hygroreceptors on antennae and body. There are sensory papillae.

ii No physiological details are known.

b Hormones are not yet described.

Respiration

a The respiratory organs are like those of insects, narrow diameter tracheal tubes, which are unbranched. They arise at surface pits or stigmata, scattered over the body, and form bunches ramifying into the tissues. There is a chitinous lining. No sphincters or spiracles are present.

b Because closure of the stigmata is not possible, except briefly perhaps during movements, water loss from the tracheae is uncontrolled and considerable. Gaseous exchange is probably similar to that occurring in the tracheal systems of insects.

Circulation/coelom

a There is an elongate tubular heart within a pericardial sinus. Blood enters via valved ostia. The body cavity is a haemocoel sub-divided into sinuses.

b The heart propels blood forward where it fills the capacious haemocoel.

c Blood is colourless and some amoebocytes are present.

Reproduction

a The sexes are separate, and there is sexual dimorphism (males have fewer legs). Crural glands in each segment are found in males but not females (except *Peripatus capensis*) and may have a sexual function. Gonads are paired, in both sexes. Males form spermatophores, females possess an ovipositor (in oviparous species), and seminal receptacles.

Phylum ONYCHOPHORA

b Mating has not been observed but there must be copulation and internal fertilization (adult females always carry live sperm in the seminal receptacles). In one species sperm gain access to the haemocoel after spermatophores are inserted into the body wall of the female.

Ooperipatus lays large, shelled eggs, not retained (oviparous).

Eoperipatus has large eggs and is ovoviviparous (eggs develop in uterus).

Peripatopsis has small eggs, and retains them in the uterus where they absorb nutrient.

Peripatus forms a true placenta and has small eggs (viviparous).

Reference

Grassé P.P. (Editor) 1949. *Traité de Zoologie*, Vol. VI. Masson et Cie, Paris.

22 Phylum ARTHROPODA

Comprises about 1 million known species, the class Insecta accounting for the vast majority. Size range microscopic to massive (up to a few metres).

Characteristics

1 Phylum contains marine, fresh water and terrestrial species.
2 Triploblastic, coelomate, bilaterally symmetrical metazoans.
3 Show metameric segmentation.
4 Have paired, jointed appendages, with at least one pair of functional jaws.
5 Possess a chitinous exoskeleton (the cuticle).
6 Growth and development proceed via a series of moults and intermoult growth periods.
7 All muscles are striated, with the muscles of the body wall as discrete bundles.
8 A tubular gut proceeds from mouth to anus.
9 The CNS has a pair of anterior supra-oesophageal ganglia connected by commissures to a ventral chain of segmental ganglia.
10 Organ systems are well developed.
11 The coelom is generally reduced in extent. The body cavity is a haemocoel. A heart with ostia is usually present.
12 Cilia are lacking, except as components of sense organs.
13 The sexes are typically separate.

Larval form and metamorphosis

Because it possesses an inelastic exoskeleton an arthropod can only increase in bulk when the exoskeleton is shed, as at a moult. Growth is said to be step-wise or discontinuous, with an ecdysis occurring between growth periods. Prior to ecdysis (which is under endocrine control) much of the procuticle is absorbed by the epidermis and a new cuticle is laid down beneath the old. The new cuticle does not become hardened until after the moult. In ecdysis the cuticle of the exoskeleton is shed together with other structures of a cuticular nature, e.g. linings of tracheae, foregut and hindgut.

In the classes Pauropoda, Diplopoda and Symphyla the young which hatch from the eggs resemble the adults but have fewer segments and legs. Subsequent growth is anamorphic, i.e. the juvenile attains adult form by the addition of extra segments and legs at successive moults. In the class Chilopoda the young hatchlings either already possess the adult complement of segments and legs or if not, attain them during a period of anamorphic development.

Most members of the class Arachnida have no larva or metamorphosis. The young from the eggs closely resemble the adults and attain adult form by growth and ecdyses.

The class Merostomata has pelagic larvae, which in early development show a 'tribolite' stage with few gill-books and a short telson. The young larva resembles the adult and metamorphoses through a series of moults (up to 14 instars) with the addition of segments and elongation of the telson. The class Pycnogonida has a protonymphon larva with a pair each of chelicerae, palps and ovigerous legs, but lacking a trunk. Development resembles that of the Diplopoda with the adult complement of trunk segments and legs being achieved after a series of moults.

In the class Crustacea there is a wide variety of larval forms (Fig. 22.1), often several per species. The larva which hatches from the egg attains adult form by a series of metamorphoses. The nauplius larva is common to many crustacean groups. It has a simple form, possesses antennules, antennae, mandibles and a median eye and the thorax is not visibly segmented. It is the first larva in many groups (e.g. Copepoda, Cirripedia) and is sometimes succeeded by the slightly more developed metanauplius (e.g. Branchiopoda, some Decapoda). However the metanauplius itself may be the first free larva (e.g. Cephalocarida, Mystacocarida). The names and body forms of succeeding larval stages vary with the group.

In the class Insecta the intermoult periods are called stadia and the insect during a stadium is known as an instar. The Apterygota or primarily wingless insects do not show metamorphosis. The juveniles resemble adults at hatching and attain adult size and maturity by a series of moults.

In the Pterygota (winged insects) the larva which hatches from the egg is wingless and lacks sexual appendages. It undergoes metamorphosis to achieve the adult form. In the Exopterygota (Hemimetabola) the young stages, or nymphs, are similar in body form to the adults and have wing rudiments as visible small projections on the outside of the body. Metamorphosis is a less drastic process in these insects than in the Endopterygota or Holometabola whose larvae are very dissimilar to the adults and may occupy different habitats. Their wings develop internally and are not visible during larval stages. Metamorphosis is complex with an intermediate stage, the pupa, between larva and adult. The pupa is often apparently inactive; it does not feed and it may be protected by a cocoon or puparium or some other device. Within the pupa many of the larval systems disintegrate and adult tissues and organs are built up from buds or imaginal discs.

Adult body form

The arthropod body is metamerically segmented with, primitively, serial repetition of appendages, internal organs, muscles, nerves, etc. However in more highly-developed forms there has been a tendency for tagmatization to occur with segments becoming organized into groups or tagmata specialized to perform particular functions. The composition of tagmata varies in the major groups. There is a head or definite anterior end which bears the mouth and sense organs.

The appendages are jointed and are present in varying numbers and forms in the different classes. They may be locomotory or modified as mouthparts or as genital, sensory, respiratory structures etc.

Arthropods possess a chitinous exoskeleton. The epidermis (beneath which is a basement membrane) secretes the overlying cuticle (Fig. 22.2). Immediately above the epidermis is the more massive procuticle responsible for the skeletal function. It is composed of chitin and proteins and in this form (flexible but non-elastic) it is present at joints and in larvae. Normally it becomes hardened by sclerotization or tanning of the protein

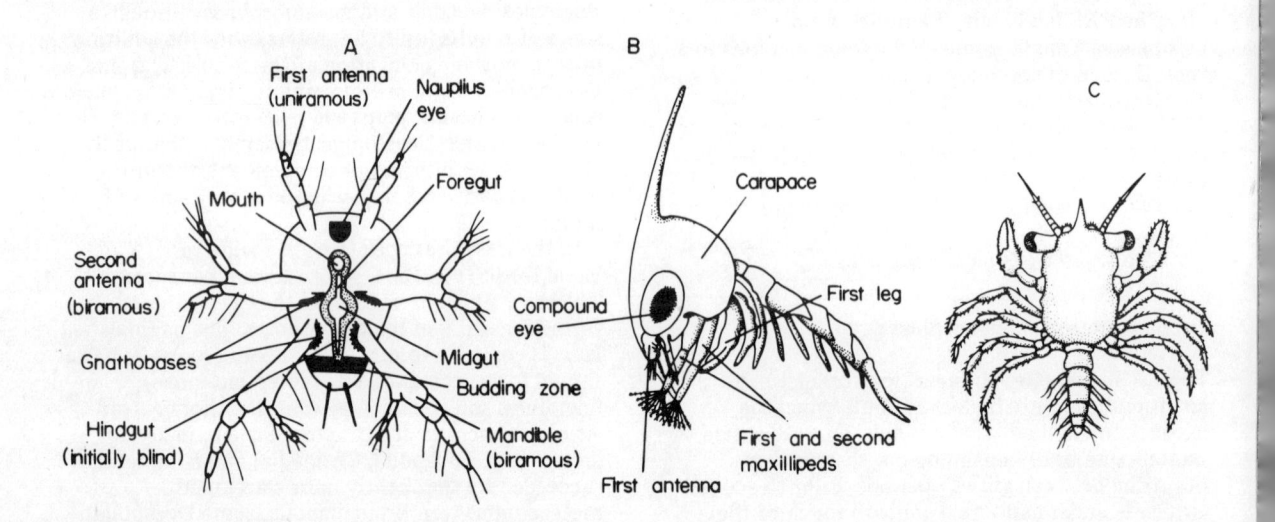

Fig. 22.1 Three crustacean larval forms. A. Nauplius (after Russell-Hunter). B. Zoea (after Lebour). C. Megalopa (after Bate).

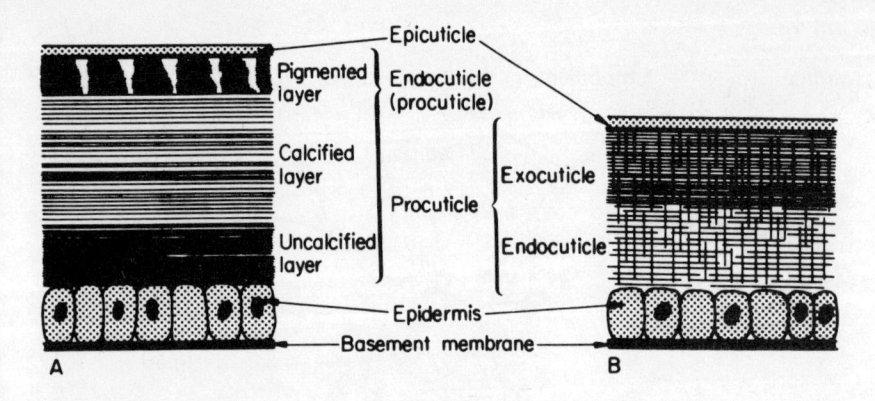

Fig. 22.2 Structure of the arthropod integument. A. In decapod crustaceans. B. In insects (A and B after Russell-Hunter).

component (as in insects, but occurs to some extent in all arthropods) or by calcification (as in crustaceans and some diplopods). It occurs as plates (e.g. dorsal tergites and ventral sternites). Overlying the procuticle is a thin epicuticle which incorporates layers of wax molecules and it is this layer which is responsible for any non-wettability and impermeability.

Frequently extensions of the exoskeleton penetrate deep into the body forming (i) attachment places (apodemes) for muscles and (ii) a type of endoskeleton (endophragmal).

1 *Class* TRILOBITA   no living species.

2 *Class* MEROSTOMATA   e.g. *Limulus, Tachypleus*. The living species are the horseshoe crabs or Xiphosura (Fig. 22.3). Generally marine, they resemble Arachnida, with 3 tagmata: (1) a prosoma with a broad shield-shaped carapace, a pair of chelicerae and 5 similar pairs of ambulatory legs (all but the last pair, which is spinous, bear distal chelae and between the basal parts of the legs are spiny gnathobases) and an opisthosoma joined by a hinge to the prosoma and composed of (2) an anterior mesosoma of 6 segments with a dorsal carapace and (3) a posterior metasoma of 3 fused segments and terminal telson. The first pair of mesosomal appendages form a genital operculum, the others are flattened, expanded and biramous with the lateral branch carrying gill-books. The metasoma lacks appendages.

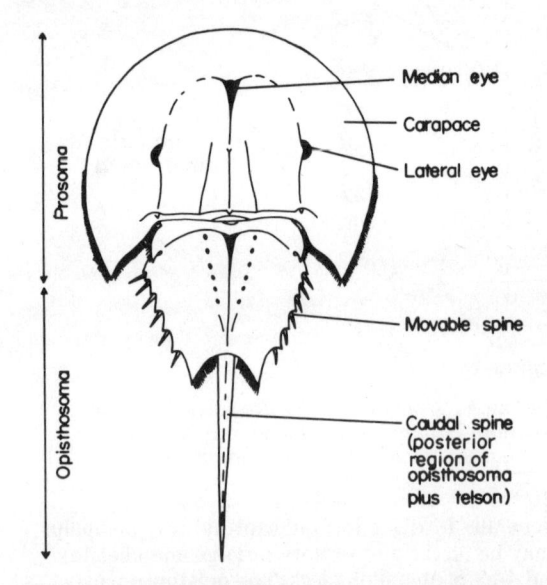

Fig. 22.3 Merostomata. Dorsal view of *Limulus* (after Snow).

3 *Class* ARACHNIDA   scorpions, spiders, ticks, mites etc. (Fig. 22.4), e.g. *Androctonus, Araneus, Ixodes*.
Terrestrial; body of anterior prosoma and an abdomen or opisthosoma, with segmental fusion in both. The prosoma (developmentally of 7 segments) bears 6 pairs of appendages: the pre-oral chelicerae

Phylum ARTHROPODA   107

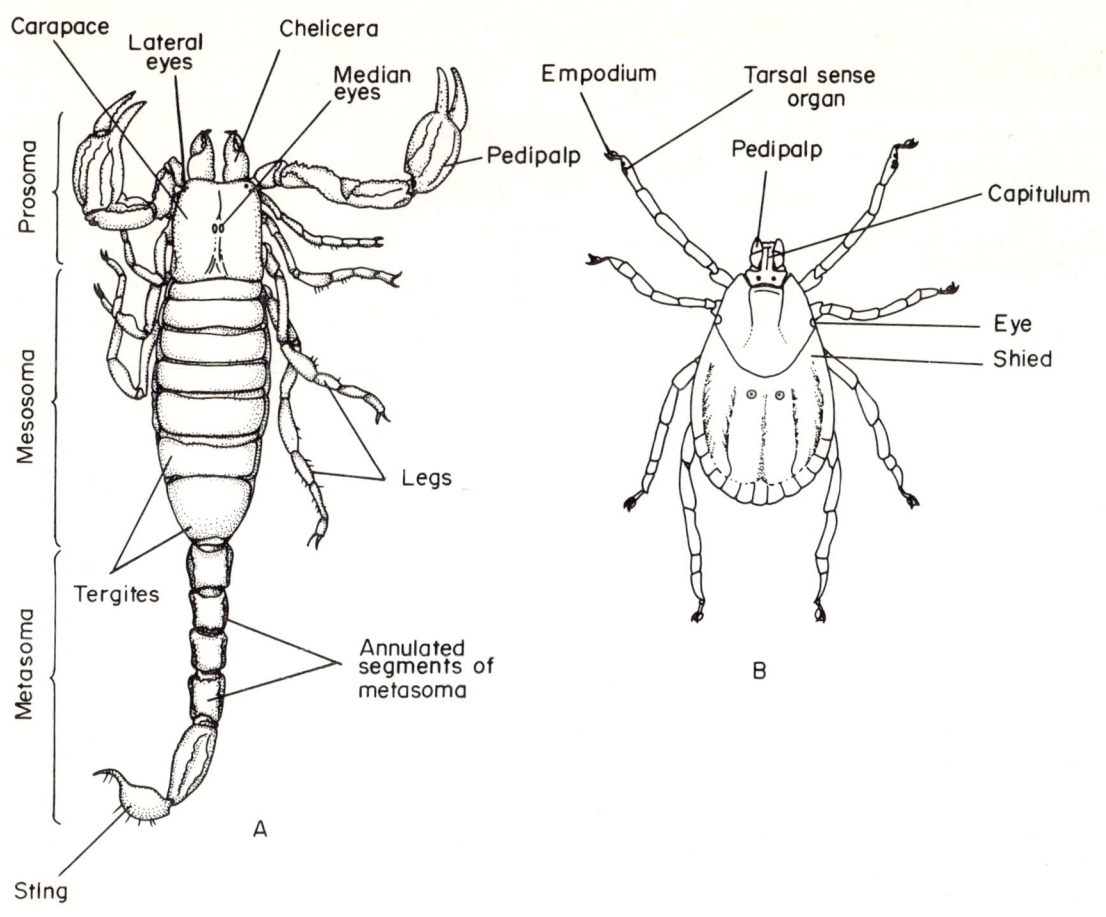

Fig. 22.4 Arachnida. A. Dorsal view of the scorpion *Urodacus* (after Main). B. The tick *Dermacentor variabilis* (after Snodgrass).

(generally involved in food capture), the pedipalps (may be tactile and sensory or large and chelate) and 4 pairs of walking legs. The opisthosoma is generally of 12 fused segments (a pre-abdomen of 7 and a post-abdomen of 5 in scorpions), and lacks appendages except for the pectines of scorpions and the spinnerets of spiders. Legs possess only flexor muscles; extension is by hydrostatic pressure.

4 *Class* PYCNOGONIDA sea-spiders (Fig. 22.5), e.g. *Nymphon, Pycnogonum*.
Marine; head bears a proboscis, a pair of chelicerae and a pair of palps; possess a greatly reduced trunk with 4 pairs of enlarged walking legs, and a vestigial abdomen; may be related to both arachnids and crustaceans.

5 *Class* CRUSTACEA includes lobsters, crabs, shrimps, barnacles etc., e.g. *Artemia, Daphnia, Cypris, Calanus, Squilla, Mysis, Cancer, Homarus*. Primarily aquatic, most species marine but with some fresh-water and terrestrial forms; morphologically very diverse (Fig. 22.6); possess a head of 6 segments fused with the acron, a thorax, and an abdomen with a post-segmental telson; the total number of segments varies; some thoracic

108 Phylum ARTHROPODA

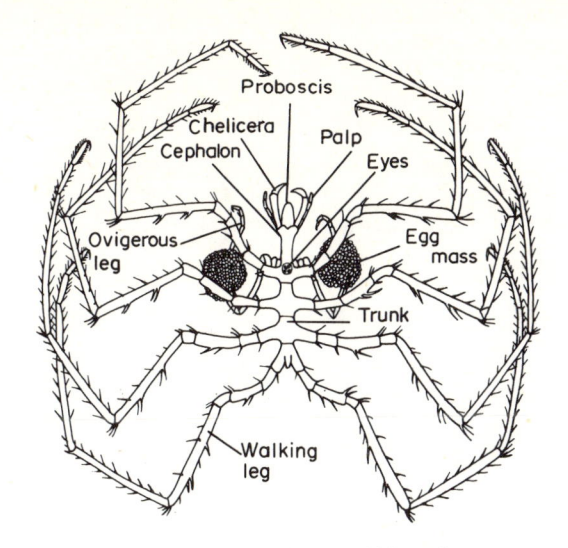

Fig. 22.5 Pycnogonida. Male of *Nymphon rubrum* (after Sars).

segments may be fused with the head to form a cephalothorax, and a carapace is found in some orders; head appendages (see Fig. 22.7) are as shown below.

1st antennae or antennules	on 2nd metamere
*2nd antennae	on 3rd metamere (*unique to Crustacea)
the mandibles	on 4th metamere
the 1st maxillae or maxillules	on 5th metamere
the 2nd maxillae	on 6th metamere.

Thoracic and abdominal appendages may be variously modified for feeding, locomotion (swimming or walking), respiration or as accessory reproductive organs (Fig. 22.7). Except for the antennules, the appendages are typically biramous (Fig. 22.8); the basal protopodite bears two arms, an inner endopodite and an outer exopodite. The exoskeleton is calcified, composed of plates or sclerites and may have protrusions such as setae.

The following 4 classes are terrestrial, mandibulate, so-called myriapodous arthropods with a series of similar leg-bearing segments and show little evidence of tagmatization.

6 *Class* **DIPLOPODA** the millipedes, e.g. *Glomeris*, *Polydesmus*.
Head with a pair each of antennae and mandibles,

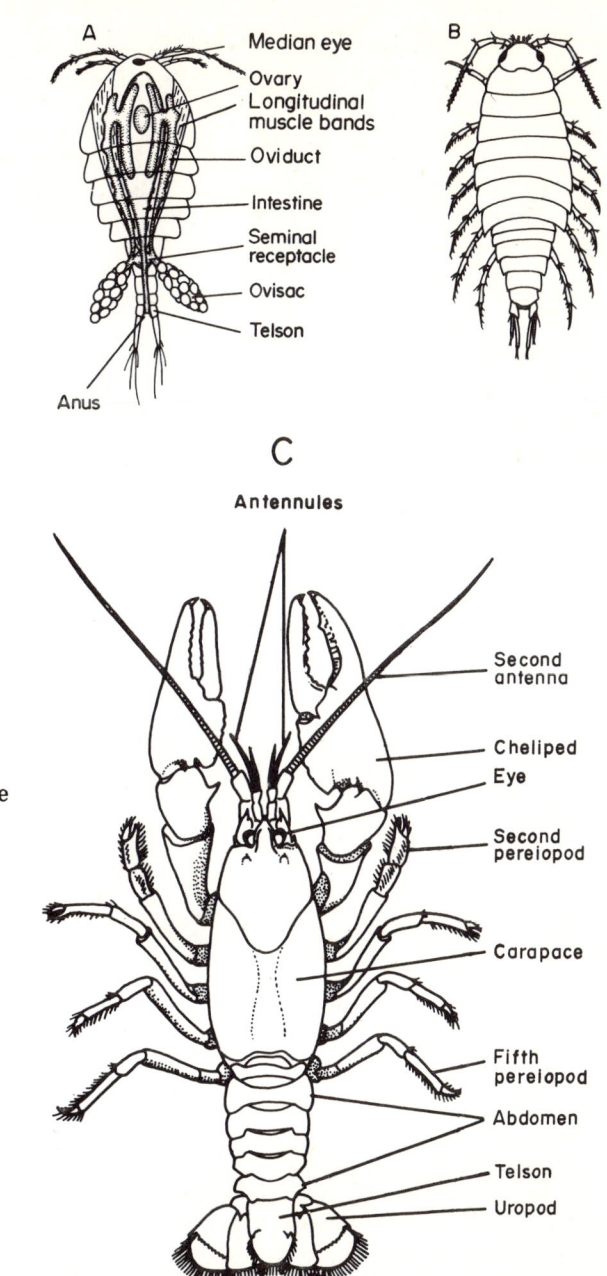

Fig. 22.6 Crustacea. A. Schematic representation of the copepod *Cyclops* (after Bullough). B. *Ligidium*, a terrestrial isopod (after Walker). C. Male of the decapod *Cambarus longulus* (after Snodgrass).

Phylum ARTHROPODA 109

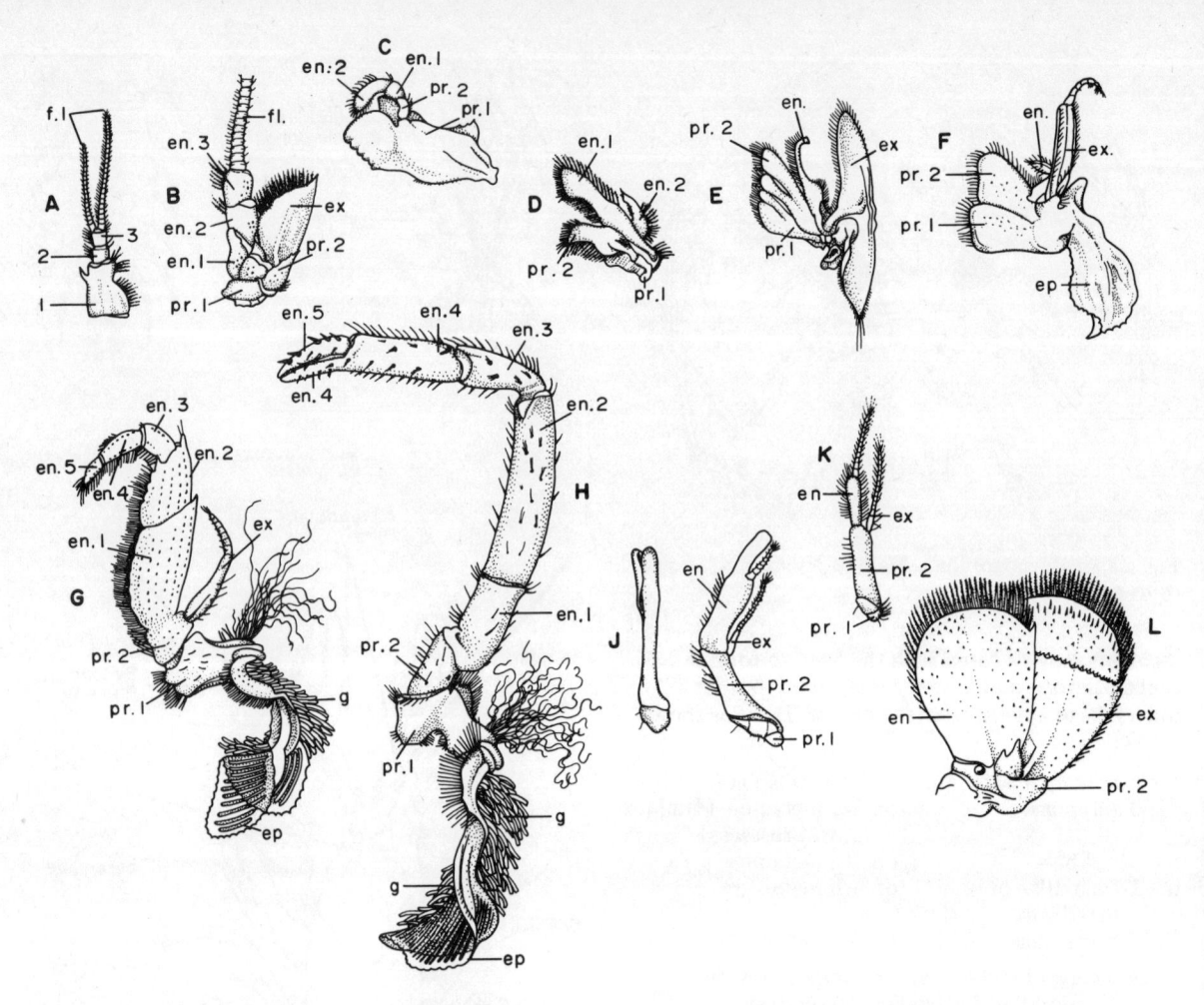

Fig. 22.7 Some appendages of the decapod crustacean *Astacus*. A. 1st antenna or antennule. B. 2nd antenna. C. Mandible. D. 1st maxilla. E. 2nd maxilla. F. 1st maxilliped. G. 3rd maxilliped. H. 3rd walking leg. J. Copulatory organs. K. Pleopod. L. Uropod (A—L after Huxley). (en endopodite; en. 1—5, podomeres of endopodite; ep, epipodite; ex, exopodite; fl, flagellum; g, gill; pr.1, pr.2, podomeres of protopodite; 1—3, podomeres of axis of antennule).

and with the fused 1st maxillae forming a gnathochilarium; the first trunk segment is the collum, followed by a number of leg-bearing segments (mostly diplosegments with two pairs of legs), then a number of apodous segments (without legs) and a telson.

7 *Class* PAUROPODA e.g. *Pauropus*.
Minute, soft-bodied; head with a pair each of antennae (each of 3 separate flagella), mandibles and first maxillae; trunk of 12 segments, the first being the collum; collum and segments 11 and 12 lack legs, the rest are podous.

8 *Class* CHILOPODA the centipedes (Fig. 22.9), e.g. *Geophilus*, *Scutigera*.
Head with a pair each of filiform antennae, mandibles, first and second maxillae (maxillae may

110   Phylum ARTHROPODA

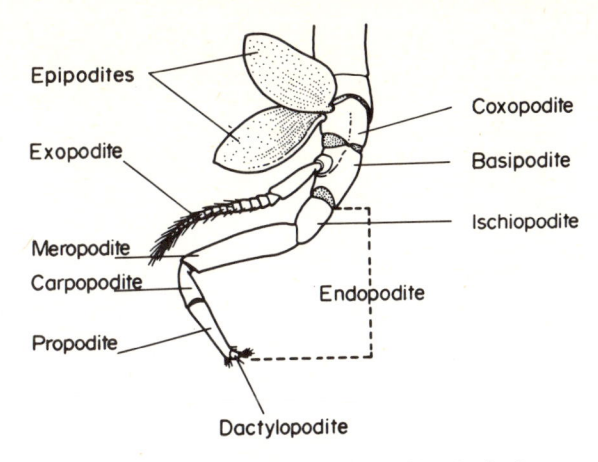

Fig. 22.8 Third left pereiopod (walking leg) of *Anaspides tasmaniae*, showing the biramous condition (after Snodgrass).

be united medially); 1st trunk segment with modified appendages, the poison claws; remaining segments modified differently with order; several terminal segments without legs.

9 *Class* SYMPHYLA   e.g. *Scutigerella, Symphylella*.
With resemblances to both centipedes and apterygote insects; head with antennae, mandibles, 1st and 2nd maxillae (2nd maxillae united medially to form a labium); trunk with 14 segments, the first 12 of which bear legs, the thirteenth a pair of spinnerets and the fourteenth a pair of sensory hairs; most of the podous segments also have a pair of styles and a pair of eversible sacs.

10 *Class* INSECTA   the insects (Fig. 22.10), e.g. *Lepisma* (Apterygota), *Aeschna, Periplaneta, Forficula, Locusta, Pediculus, Papilio, Glossina, Pulex, Apis, Dytiscus*.
Body with well-defined head, thorax and abdomen which may be separated by constrictions; head of a number (probably 6) of fused segments; thorax of 3 segments and abdomen of 9—11 segments and a vestigial telson; head appendages are a pair each of antennae, mandibles, 1st maxillae and 2nd maxillae (fused medially to form a labium); each thoracic segment bears a pair of legs (Fig. 22.10C) and in pterygote insects a pair of wings is borne on each of the 2nd and 3rd thoracic segments (one or both pairs lost in some insects or non-functional in

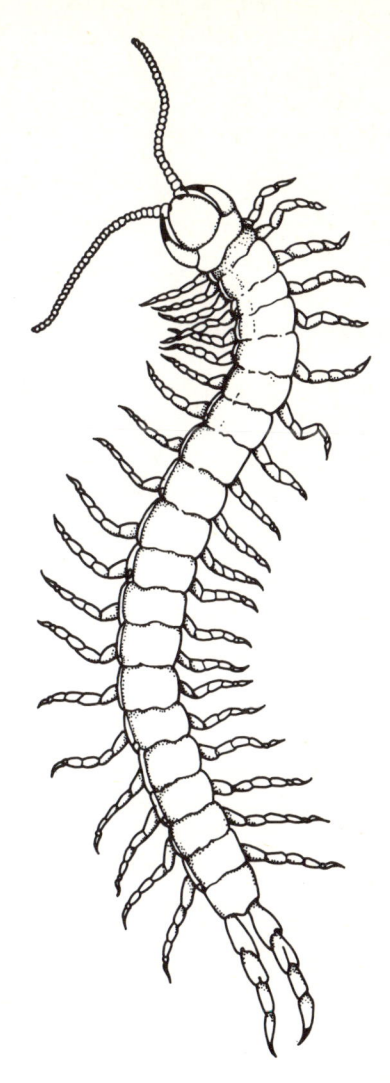

Fig. 22.9 Chilopoda. The centipede *Scolopendra* (after Koch).

flight); except in the apterygote insects there are no abdominal appendages anterior to the genital segments; frequently bear appendages on the genital segments; appendages (where present) on the 11th segment called cerci; external bodily form varies greatly in the different orders; where soft-bodied (e.g. caterpillars) the unsclerotized cuticle is held taut by internal pressure of the blood (hydrostatic skeleton).

Phylum ARTHROPODA   111

Feeding

Table 22.1

Class	Type of feeding	Food
Merostomata	Scavenging, carnivorous	Molluscs, worms, algae
Arachnida	Predatory, carnivorous, chelicerate	Invertebrates
Pycnogonida	Predatory, carnivorous, suctorial	Sessile organisms, e.g. sponges, anemones
Crustacea	(i) Filter-feeding using cirri (modified limbs)	Plankton, detritus
	(ii) Macrophagous, scavenging	
	(iii) Predatory, carnivorous	Invertebrates, fish
	(iv) Detrital feeding (omnivorous)	
	(v) Parasitic	Fluids, blood, absorbed protein
Diplopoda	Mostly saprovorous	Ingest soil & decaying vegetation
Pauropoda	Scavenging	Scavenging in soil litter
Chilopoda	Predatory, carnivorous	Worms, molluscs, other arthropods
Symphyla	Mostly saprovorous	Decaying vegetation
Insecta	(i) Herbivorous	All parts of living plants
	(ii) Saprovorous	Decaying vegetation
	(iii) Predatory, carnivorous	Smaller insects or other animals
	(iv) Feeding on fungi and bacteria	
	(v) Filter-feeding (e.g. mosquito larvae)	Suspended particles
	(vi) Feeding on dung, carrion and blood	

The arthropod gut is tubular and divisible into cuticle-lined stomodaeum (foregut) and proctodaeum (hindgut) (Fig. 22.11) and a generally non-cuticularized midgut. There is usually a valve separating fore- from midgut and mid- from hindgut. The midgut frequently bears caeca or diverticula. Near the junction of midgut and hindgut Malpighian tubules may be present. The anus is terminal.

Class Merostomata. The chelae are used for food capture. The food is crushed between the gnathobases and pushed into the mouth between the 1st pair of legs. The gizzard triturates the food,

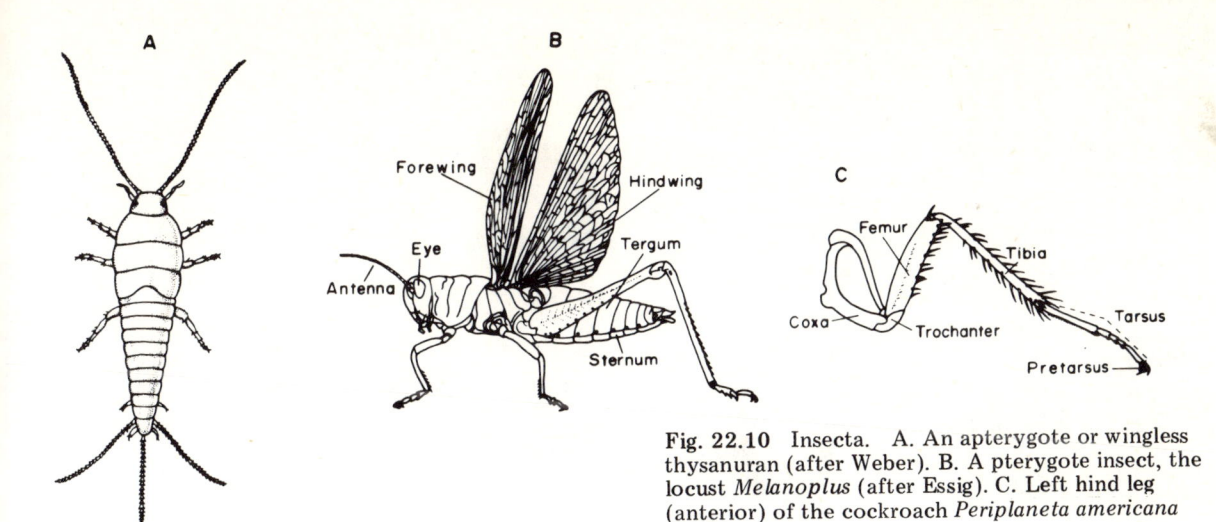

Fig. 22.10 Insecta. A. An apterygote or wingless thysanuran (after Weber). B. A pterygote insect, the locust *Melanoplus* (after Essig). C. Left hind leg (anterior) of the cockroach *Periplaneta americana* (after Snodgrass).

solid particles being ejected from the mouth and fluid passing through a valve to the midgut. Intracellular digestion occurs in diverticula of the midgut, the digestive glands.

Class Arachnida. Chelicerae (Fig. 22.12A,B) and pedipalps may be modified for prey capture. The mouthparts are generally modified for the intake of liquid food, forming a pre-oral cavity which may bear hairs that act as a filter. The mouth leads into a sucking pharynx, followed by a narrow oesophagus and a pumping stomach. Maxillary salivary glands may open into the foregut. A valve separates oesophagus and gizzard. Diverticula of the midgut, lying in the prosoma, secrete enzymes which reduce the food to an absorbable form, while branched diverticula (forming the chylenteron or liver) lying in the opisthosoma are responsible for absorption and storage.

Class Pycnogonida. Pycnogonids also feed on fluids. The mouth lies at the tip of a proboscis, in front of a sucking pharynx. A chitinous filter prevents entry of solid material. The gut is simple, the stomach small and the digestive diverticula extend into the legs. Digestion and absorption occur in the midgut.

Class Crustacea. In addition to mandibles and 1st and 2nd maxillae other structures which may be involved in feeding are the labrum, labium, antennae, antennules and thoracic appendages. Filter-feeders may utilize thoracic appendages (Fig. 22.12C), maxillae, mandibles, antennules or antennae. Crustaceans feeding on large particles, and predatory species, have the limbs modified for grasping (chelae, sub-chelae), crushing (spines, gnathobases), catching (maxillipeds, chelae), and tearing prey apart.

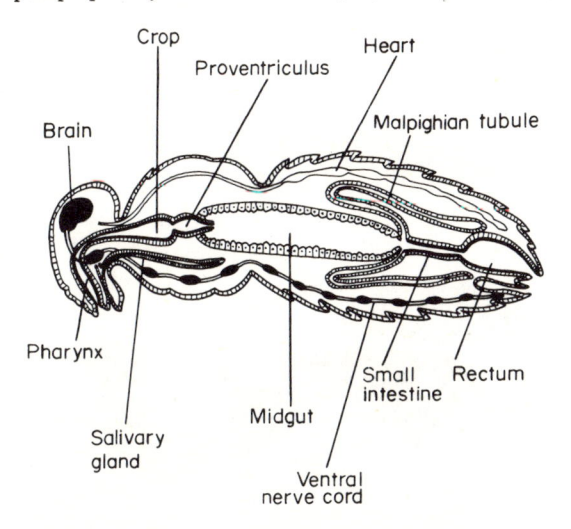

Fig. 22.11 Diagram of the alimentary canal of an insect showing the cuticular lining of fore- and hindgut (after Snodgrass). Cuticle is indicated by heavy line in fore and hind gut.

Phylum ARTHROPODA   113

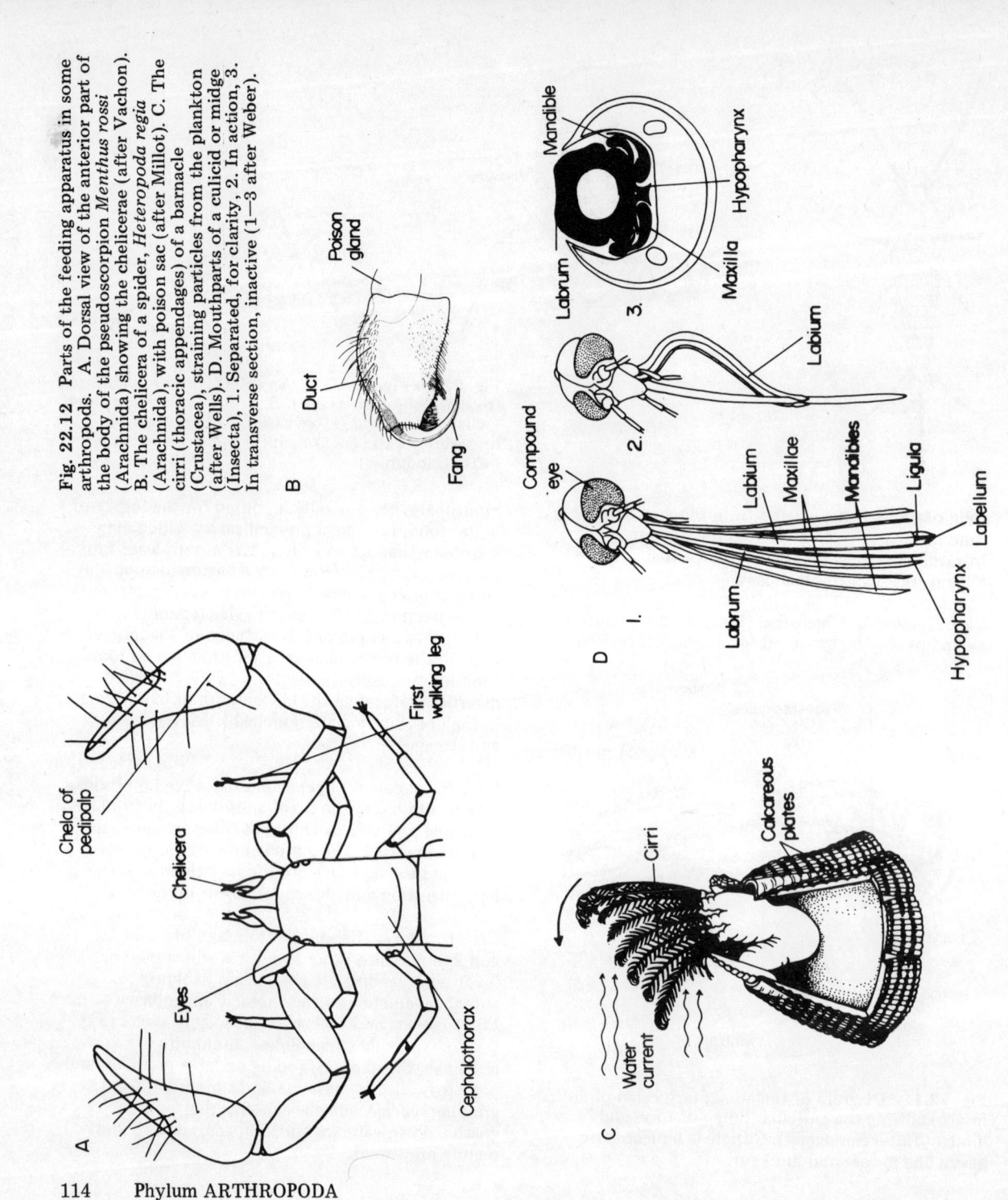

Fig. 22.12 Parts of the feeding apparatus in some arthropods. A. Dorsal view of the anterior part of the body of the pseudoscorpion *Menthus rossi* (Arachnida) showing the chelicerae (after Vachon). B. The chelicera of a spider, *Heteropoda regia* (Arachnida), with poison sac (after Millot). C. The cirri (thoracic appendages) of a barnacle (Crustacea), straining particles from the plankton (after Wells). D. Mouthparts of a culicid or midge (Insecta), 1. Separated, for clarity, 2. In action, 3. In transverse section, inactive (1–3 after Weber).

114 Phylum ARTHROPODA

The mouth is ventral, the gut usually tubular. In the Malacostraca the posterior part of the foregut is armed with cuticular masticatory processes (ossicles) and filtering devices for particle sorting. Anterior, blind diverticula of the midgut (forming the hepatopancreas) are responsible for enzyme secretion, absorption and storage. Digestion is extracellular.

In the four myriapodous classes the gut is usually a straight tube. The prehensile poison claws of the class Chilopoda are used to capture and immobolize prey. The Diplopoda have large mandibles and a gnathochilarium for chewing. There may also be specializations for mucus secretion from salivary glands, for suctorial feeding or for sucking plant fluids via a 'beak'. The midgut of diplopods and chilopods lacks digestive glands. In the class Symphyla mandibles, maxillae and labrum are involved in ingestion. Details of digestive physiology are few.

Class Insecta. The structure of the mouthparts varies with mode of feeding and may be modified e.g. for biting and chewing, piercing and sucking (Fig. 22.12D) etc. In mosquito larvae (filter-feeders) long tufts of hairs on the head create the current driving food particles to the mouth and bristles on mandibles and maxillae act as a filter.

The foregut may possess a heavily muscled pharynx leading from the mouth, followed by a narrow oesophagus which may be expanded posteriorly as a crop for food storage. Behind this and leading to the midgut is the proventriculus, which may function simply as a valve or be armed with cuticular ridges and have a triturating function. The proventricular valve separates fore- and midgut. The midgut often produces a thin, cuticular peritrophic membrane (anteriorly or from the general midgut wall) which protects the midgut epithelium and is voided wrapped around the faeces. The midgut often bears caeca anteriorly.

In the crop and gizzard food may be exposed to enzymes from salivary glands and midgut. Most digestion (extracellular) and absorption occur in the midgut. In some predaceous beetles (e.g. *Dytiscus* larvae) some extra-corporeal digestion may occur with the fluid results being ingested.

Between midgut and hindgut there is often a pyloric valve. The anterior hindgut may be divisible into an ileum and colon or may be a simple tube leading to the rectum and anus.

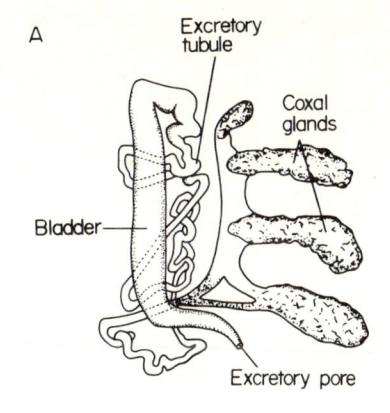

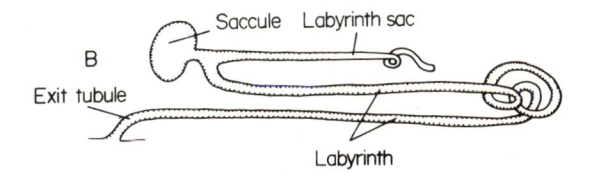

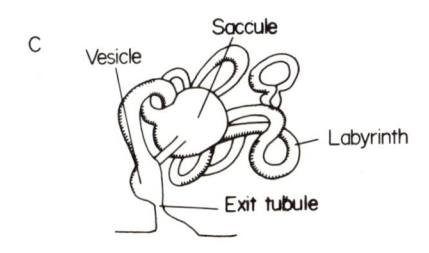

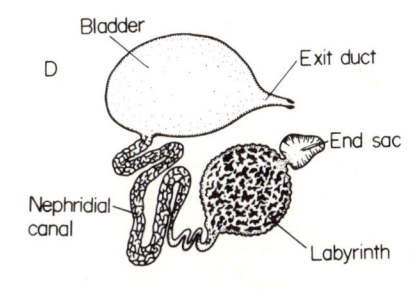

Fig. 22.13 Some arthropod excretory organs. A. Coxal glands and ducts of *Limulus* (Merostomata) (after Patten and Hazen). B. Coxal gland of Solifugae (Arachnida). C. Coxal gland of Scorpionida (Arachnida) (B and C after Buxton). D. Antennary or green gland of *Astacus* (Crustacea), unravelled (after Marchal).

Phylum ARTHROPODA

Osmoregulation/excretion

a See Table 22.2 below

Table 22.2 Arthropod excretory organs

Class	Excretory organs	Structure	Excretory products
Merostomata	Coxal glands (segments 2–5) (Fig. 22.13A)	Gland, duct, common coiled labyrinth leading into duct opening at base of 5th leg	Details unknown
Arachnida	Malpighian tubules	Blind tubules emptying into gut near mid- and hindgut junction	Mainly guanine
	Storage cells (e.g. in chylenteron, hypodermis)		
	Nephrocytes (in haemocoel)		
	Coxal glands (Fig. 22.13B,C)	See above, open behind coxae on segment 2, 3 or 5 of prosoma	
Pycnogonida	No special organs		
Crustacea	*Generally* a pair of green glands (2nd antennal segment) (Fig. 22.13D) *or* a pair of maxillary glands (2nd maxillary segment)	A blind sac in the haemocoel connects to a distal bladder opening at a pore; the connecting canal may be larger and more convoluted in fresh-water forms	Mainly ammonia
			Some amino nitrogen
			Some urea in fresh-water forms
			Uric acid in terrestrial isopods
	Gills, gut and nephrocytic cells may also be involved		
Diplopoda	1 or 2 pairs Malpighian tubules		Physiology unknown
Pauropoda	No organs described		Physiology unknown
Chilopoda	1 or 2 pairs Malpighian tubules		Physiology unknown
Symphyla	No organs described		Physiology unknown
Insecta	Two to many Malpighian tubules	See above	Mainly uric acid
			Urea may be present in small amounts
	Midgut and fat body may also be involved	Fat body contains 'urate' storage cells	Ammonia in some insects

b

Class Merostomata. Little information is available on the osmoregulatory abilities of this group.

Class Arachnida. A dorsal stercoral pouch of the gut, behind the Malpighian tubules, retains guanine before evacuation and further resorption of water may occur. Guanine may also be retained in storage cells.

Class Pycnogonida. Osmoregulation and excretion occur across the general body surface.

Class Crustacea. Removal of nitrogenous excretory compounds apparently does not depend primarily on the so-called excretory organs, but has been shown to be due to the permeability of the body surface, especially the gills. Osmoregulation is effected chiefly by the gills and hindgut.

Classes Diplopoda and Chilopoda. Diplopods are able to regulate water loss through the spiracles. Osmoregulation and excretion in both classes are probably similar to the processes in insects.

Class Insecta. Generally the distal portions of the Malpighian tubules permit entry of waste products from the blood. Selective resorption of water, ions and nutrients occurs at the lower end. The main nitrogenous waste product is uric acid, which has a very low solubility in water and is excreted in solid form. Uric acid passes into the gut to be voided with undigested food wastes. Further resorption of water from the faecal pellets may occur in the hindgut (e.g. by the activity of rectal pads or papillae).

In most insects specialized cells of the fat body store uric acid. In forms which lack Malpighian tubules this may be the only method of removing nitrogenous waste from circulation.

In some insects which excrete ammonia, this is produced by midgut cells, passes into the haemocoel and thence into the Malpighian tubules.

Movement

a Arthropods have been successful in all habitats and consequently the group demonstrates numerous types of movement. Movements are functionally concerned with three basic processes, namely locomotion, respiration and feeding. Sound production constitutes a minor fourth variation.

The metameric arrangement of the body of an arthropod is indicated by the repetition of the appendages concerned with these activities along the body. In some forms the same limbs have more than one function, in others there is specialization and modification to greater efficiency at one function.

1 *Locomotion.* Aquatic, terrestrial and aerial environments have been successfully colonized by various groups. The utilization of limbs for locomotion varies according to habitat, and type of locomotion.

Major **locomotory patterns**

i Swimming: widespread throughout the aquatic groups (especially Crustacea), it may be brought about by the synchronized use of a large number of similar limbs (Branchiopoda), or by certain phyllopodous limbs of one portion of the body (e.g. thorax or abdomen), or by single limbs (copepod antennae).

ii Burrowing: many arthropods can burrow, and frequently this is due to limbs and parts of the body specialized for the purpose or to the co-ordinated activity of groups of limbs (e.g. in shrimps).

iii Escape movements: result from rapid tail flips (abdominal movements) or the use of groups of limbs instead of the one or two normally involved in slow progression.

iv Walking: many groups of arthropods move by walking, either using many limbs (Diplopoda, Chilopoda) metachronously synchronized, and in differing numbers depending upon whether movement is rapid or slow (Fig. 22.14) or using a small number of limbs (8 or 10 in decapod crustaceans, 8 in arachnids, 6 in insects) for most normal progression. The legs are often all similar, though in a few cases modifications occur for feeding (e.g. xiphosuran gnathobases).

v Flight: one group, the insects, have successfully evolved flight mechanisms (Fig. 22.15) with modifications of the thoracic region, the wings; these are not limbs in the metameric series but are separately developed.

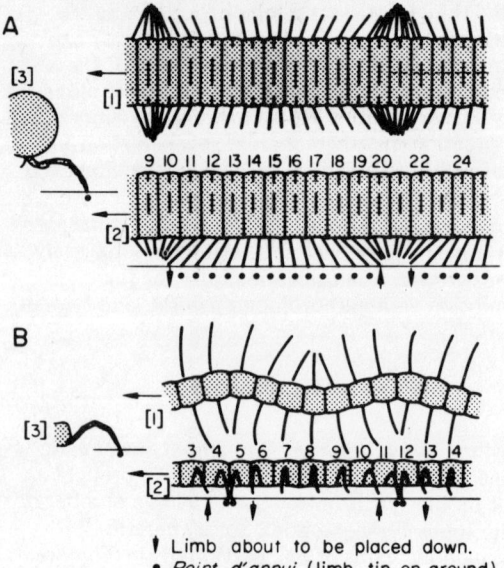

Fig. 22.14 Myriapod locomotion. A. 1. Dorsal view of a typical millipede in motion, showing 16 of the short, wide diplosegments with 32 pairs of limbs, left and right sets being in phase with each other. 2. Lateral view of the same millipede showing that at any one time most limb tips are on the ground, giving longer propulsive strokes and slow movement. 3. The diagram shows the ventral attachment of the relatively short millipede limb to the cylindrical diplosegment. B. 1. Dorsal view of a typical centipede in motion, showing 12 of the long, narrow segments each with a pair of limbs in opposite phase, and the bodily undulations which occur at speed and are mechanically inefficient. 2. Lateral view of the same showing the greater phase difference between segments, with less than a third of the limb tips on the ground at any one time, giving short swift propulsive strokes. 3. The centipede limb (*Scolopendra*) and its position of attachment (lateral) to the segment (A and B after Russell-Hunter).

2 *Respiration.* Respiration may be a consequence of

i generalized limb movement during locomotion, e.g. mysids and branchiopods;

ii utilization of specialized limbs in particular regions of the body, e.g. pleopods of Crustacea;

iii development of one or more limbs (second maxilla in decapods) as pumps to draw water through a chamber, as in many Crustacea that have developed a carapace;

iv behavioural modification (plastron respiration in aquatic insects);

v 'pumping of thoracic regions and co-ordinated spiracular opening and closing (insects).

3 *Feeding.* Obtaining food from the surroundings is an active process for arthropods and involves the activity of the limbs in a real way.

In Crustacea probably the primitive use of the limbs was in feeding and locomotion was a coincidental process. Both actions often occur simultaneously in the same limb. For example, filtration of water by copepods and by other lower Crustacea occurs during swimming movements. It may also take place, as for cirripedes, by strokes of the limbs through the ambient water (Fig. 22.12C).

The increasing cephalization of the body, however, has in many advanced cases led to modification of the anterior appendages for strictly feeding purposes.

4 *Sound production.* Noise of various frequencies is produced by stridulation, by wing rubbing, antennal scraping, joint clicking and other movements.

b The possession of a stiff exoskeleton, and the jointed nature of the limbs (Fig. 22.16) which allows flexibility, has led to the development of limbs as levers, and to the gradation of movements by specialization of bodily regions. The metameric series of limbs in arthropods has been much modified by specialization for the particular movements demanded of certain areas. Thus the cephalization of the anterior end has been followed by development of appendages organized for feeding, whilst more rearward limbs may be concerned with respiration and locomotion.

Generally the structure of a limb indicates its function; thus the larger, broader and more setose the limb the more likely it is to move water and hence to be used for swimming and feeding; the more elongate, cylindrical, rigid and stiff the limb, the more likely it is to be of value in walking, running and prey capture. Short, stubby and massive limbs are concerned in grinding and feeding;

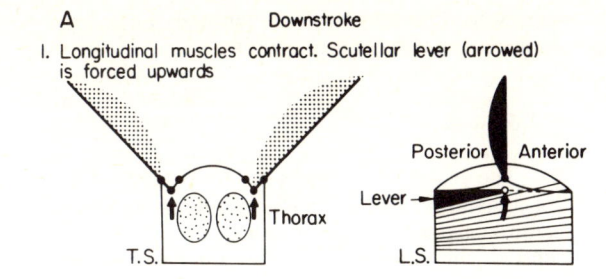

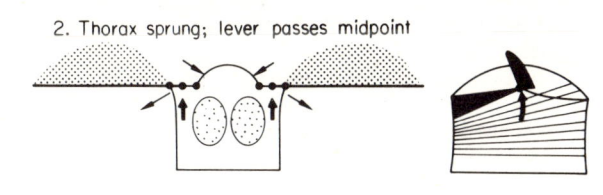

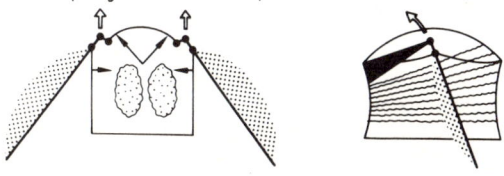

Fig. 22.15 Insect flight. This figure illustrates the 'click' mechanism of a fly's wing. Wing movement results when muscles deform the thorax. The main wing beam is pivoted on the side of the thorax and is linked at its inner end with the tip of the scutellar lever, rigidly fixed to the rear end of the thoracic box; the linkage in life is more complex than depicted here, helping to determine wing rotation as well as up and down movement. There is a second linkage between the tip of the wing beam and the top of the thorax (after Wells).

A. Downstroke. 1. Longitudinal muscles contract and the scutellar lever (arrowed) is forced upwards. 2. Thorax sprung; scutellar lever passes midpoint. 3. Sudden relaxation of tension on the longitudinal muscles as the thorax springs back into shape.
B. Upstroke. 1. Dorsoventral muscles contract and the scutellar lever (arrowed) is forced downwards. 2. Sides and top of thorax are sprung. 3. Scutellar lever passes midpoint and there is sudden removal of tension from muscles as thorax springs into shape.

extensible and tubular structures are used for sucking purposes; thin-walled and delicate appendages are often concerned in respiration.

The movement of limbs, especially for locomotion, requires considerable muscle power, and striated muscles are well developed. Muscles may end on apodemes (internal extensions of the cuticle). In some cases e.g. diplopods, chilopods and araneids, extensor muscles are lacking at most joints and leg extension is brought about by hydrostatic forces.

Length of limb gives mechanical advantage to the system so that for minimal shortening of muscle, maximum effect is gained at the peripheral end. Arthropods in general use the legs as levers, and allied to this have a rigid body axis, often reinforced by the development of a carapace or by fusion of cuticular plates. Myriapods show jointing of the body and in rapid movement flex the axis. Some have developed asymmetrically-placed scutes to minimise the deflection. Some chilopods carry out very rapid running and the legs are longer towards

Phylum ARTHROPODA

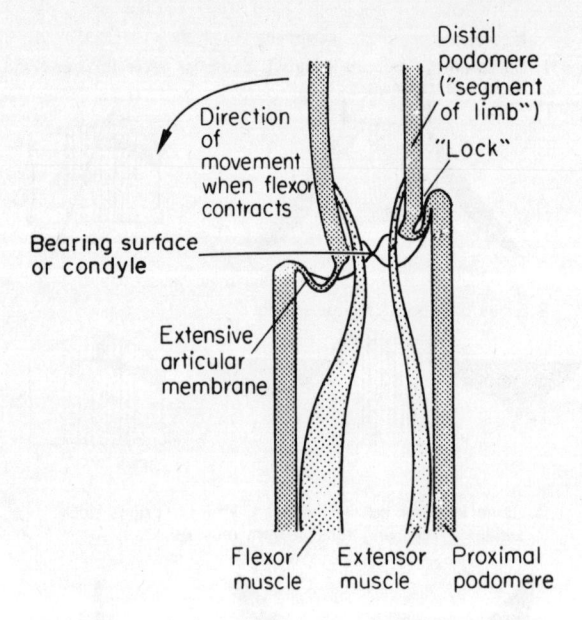

Fig. 22.16 Diagrammatic representation of a limb joint of a higher crustacean, axis of articulation perpendicular to the page (after Russell-Hunter).

the rear. Arachnids and many crustaceans have fewer legs, with some specialization for feeding, and locomotion in general is relatively slow. Insects, with only 6 legs, are capable of a range of speeds, and always have 3 legs on the ground for stability. Greater speed is achieved by raising the strike rate of the limbs.

c All arthropod muscle, including that associated with the gut, appears to be striated. Various sarcomere lengths have been described, which may be correlated with differing speeds of contraction. Many arthropods have multiple innervation of the muscle fibres and three main types of innervation are found, causing

1 fast excitation: muscle responds by rapid contraction, quick twitches and shows propagated action potentials;

2 slow excitation: muscle responds with graded potentials, and sub-maximal contractions;

3 inhibition: with consequent modification of resting muscle potentials away from their firing threshold. This may decrease the rate of contraction, or stop it altogether.

Co-ordination

a i The arthropod nervous system basically follows the annelid pattern of an anterior dorsal brain or cerebral ganglion (located near which are various sense organs), joined by two circumoesophageal connectives to a ventral nerve cord along which are segmentally-arranged ganglia. This simple basic plan may be somewhat modified when considered class by class.

The brain usually demonstrates 3 main parts: proto-, deutero- and tritocerebrum. Some 10 brain nerves are found, at least in higher Crustacea; insects and spiders have fewer. Advanced groups on the whole possess fewer nerves. A stomatogastric system supplies the gut.

The ventral nerve cord is ganglionate. Each ganglion is joined to its neighbour by interneurones, some of which may extend from one extremity of the body to the other. Giant fibres are found in some groups (e.g. orthopterans, scorpions, macruran decapods).

Types of neurones are (1) motor, large, with cell bodies in the CNS, (2) interneurones, running between all ganglia, and (3) sensory units, usually with peripheral cell bodies.

Because of the elongate, segmental nature of the arthropod body the basic pattern of the nervous system is elongate. Myriapods show about 200 ganglia, but the number is much smaller in other groups and relatively few are found in advanced crustaceans and insects. Fusion of ganglia occurs, especially in the sub-oesophageal region.

Sense organs

Many different sense organs are found in arthropods (Fig. 22.17). One may suspect that similar organs are present in all groups but this has not been adequately demonstrated. They include compound eyes (varying from those with few facets to many), ocelli, ears, setae (innervated by mechanoreceptors and chemoreceptors), statocysts (sensory elements are mechanoreceptors), proprioceptors at joint articulations or aligned with special non-postural muscles, hair plates, slit sensilla (arachnids), soft-cuticle receptors, halteres for equilibrium

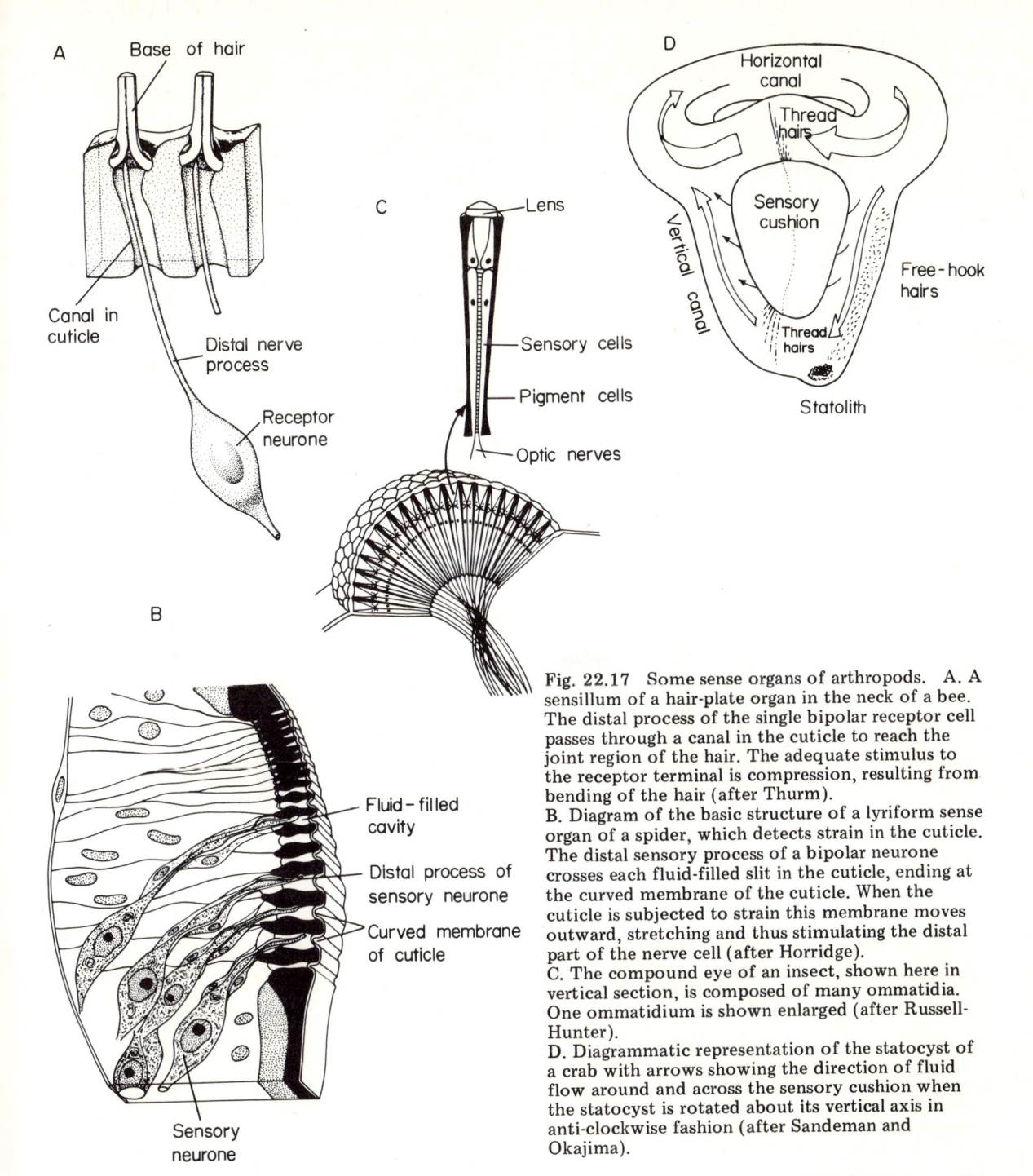

Fig. 22.17 Some sense organs of arthropods. A. A sensillum of a hair-plate organ in the neck of a bee. The distal process of the single bipolar receptor cell passes through a canal in the cuticle to reach the joint region of the hair. The adequate stimulus to the receptor terminal is compression, resulting from bending of the hair (after Thurm).
B. Diagram of the basic structure of a lyriform sense organ of a spider, which detects strain in the cuticle. The distal sensory process of a bipolar neurone crosses each fluid-filled slit in the cuticle, ending at the curved membrane of the cuticle. When the cuticle is subjected to strain this membrane moves outward, stretching and thus stimulating the distal part of the nerve cell (after Horridge).
C. The compound eye of an insect, shown here in vertical section, is composed of many ommatidia. One ommatidium is shown enlarged (after Russell-Hunter).
D. Diagrammatic representation of the statocyst of a crab with arrows showing the direction of fluid flow around and across the sensory cushion when the statocyst is rotated about its vertical axis in anti-clockwise fashion (after Sandeman and Okajima).

Phylum ARTHROPODA

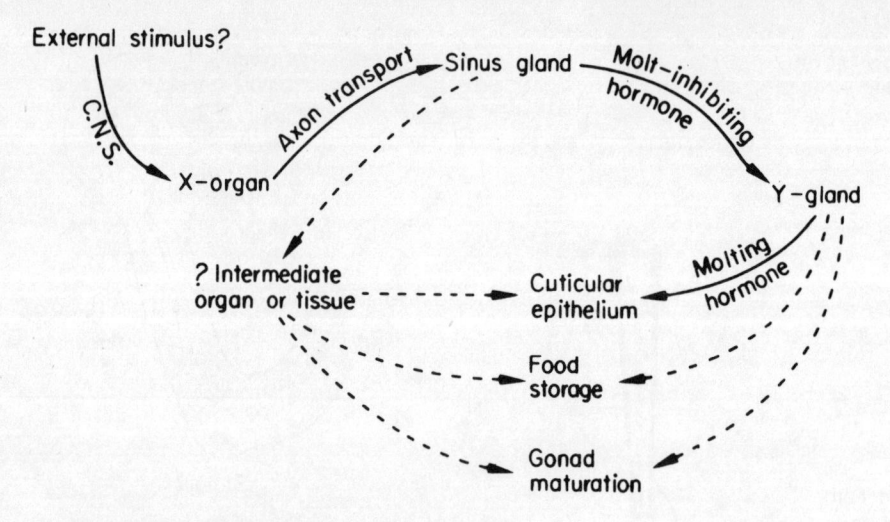

Fig. 22.18 An outline of neuroendocrine interactions controlling moulting in higher crustaceans (after Russell-Hunter).

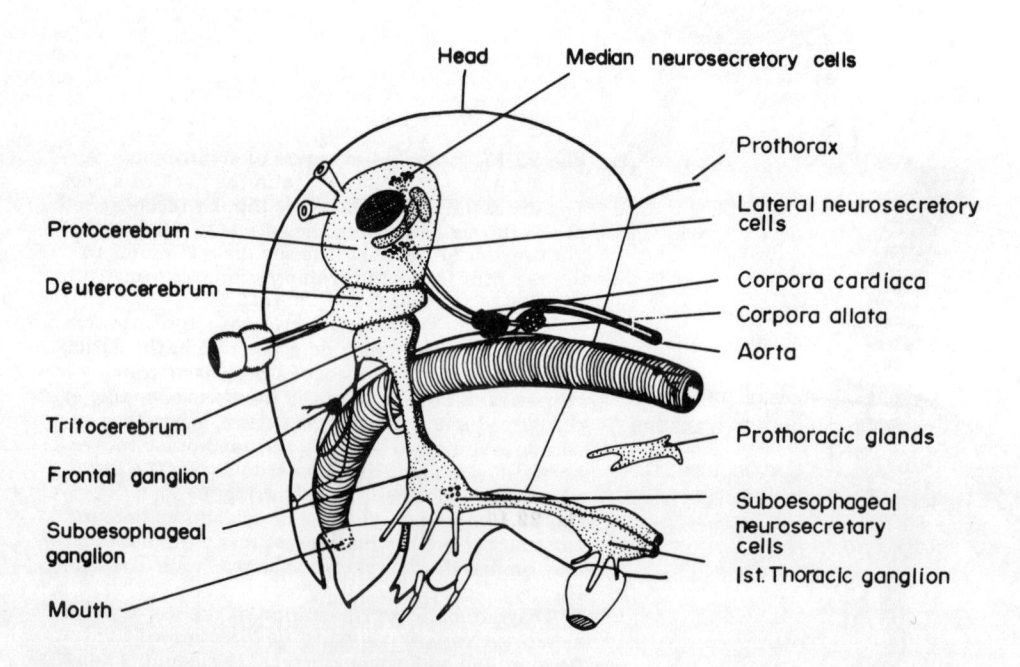

Fig. 22.19 Lateral view of the central nervous system and sites of hormone production in the head and thorax of an insect (after Jenkin).

122   Phylum ARTHROPODA

(Diptera). Eyes are rhabdomeric in structure. Proprioceptors may be bipolar or multipolar. Some receptors are formed from ciliary dendritic endings.

ii Much work has been carried out in the last two decades on the nervous physiology of arthropods. Details are available (in particular for Crustacea, Insecta, Merostomata) of microanatomy, dendritic fields, axonal pathways, transmitters, neuromuscular interactions, inhibition, receptor potentials, oscillator neurones, rhythmic pacemakers, interneurones, motor units, command fibres, patterned output, sensory responses and other aspects. It is impossible to summarize these findings in a few lines and the reader is referred to Bullock & Horridge (see Refs.).

b *Hormones* See Table 22.3 below.

Fig. 22.19 shows some of the organs in insects which are concerned in hormone production.

Table 22.3 Arthropod hormones — the processes influenced and the sites of hormone production.

Process	Merostomata	Arachnida	Pycnogonida	Myriapoda	Crustacea	Insecta
Moult (1) inhibition		protocerebrum Schneider's organ		protocerebrum	x organ	corpora allata (neotenin)
(2) promotion		protocerebrum		protocerebrum	y organ (see Fig. 22.18) crustecdysone	tentorial glands, protocerebrum, prothoracic glands, (ecdysone)
Colour change	nerve cord				sinus gland, post-commissure organ	corpora allata; neurosecretion
Water balance					sinus gland x organ	cerebral ganglia and ventral nerve cord
Retinal pigment migration					x organ	
Carbohydrate metabolism					x organ eyestalks	corpora cardiaca
Heart rate			Sokolow's organ		pericardial organs	corpora cardiaca
Sexual maturation				CNS	androgenic gland	pars intercerebralis corpora allata
Diapause						sub-oesophageal ganglion
Behaviour					pheromones in females	pheromones well represented in several groups
Lipid metabolism						corpora cardiaca

Phylum ARTHROPODA 123

Respiration

a See Table 22.4 below.

Table 22.4 Arthropod respiratory organs.

Class	Organs	Structure
Merostomata	Gill-books on all but 1st pair of opisthosomal appendages	Lamellate, with thin-walled leaflets protected by lateral extension of abdominal carapace and ventral operculum
Arachnida	(i) Lung-books in pairs at bases of rudimentary abdominal appendages	Maximum number four pairs; cuticular sac-like invaginations with walls thrown into numerous folds or leaves, stacked at free ends; cavity opens externally by a spiracle
	and/or (ii) Tracheae	Tubular cuticular invaginations as in insects, opening externally by spiracles
Crustacea	(i) Gills borne on thoracic or abdominal appendages	Thin-walled plate-like or feathery outgrowths
	(ii) Lungs in abdominal pleopods or under carapace	Lungs sac-like, occurring in terrestrial forms
Diplopoda	Tracheae	2 pairs of spiracles per diplosegment
Chilopoda	Tracheae	Spiracles variable in number and arrangement
Symphyla	Tracheae	1 pair of spiracles on head open into a pair of tracheae extending to first 3 segments only
Insecta	(i) Tracheae	Spiracles (primitively 1 pair/segment) open into tracheae which branch and anastomose with others; whole system in continuity; 2 main longitudinal trunks; tracheae prevented from collapsing by a lining of spiral cuticular ridges; tracheae branch to supply all tissues, and terminate in tracheoles
	(ii) Tracheal gills in some aquatic larvae	Occur externally or as rectal evaginations, well supplied with tracheae
	(iii) Tracheae + air-sacs	The air-sacs are much-expanded thin-walled tracheae, which form air reserves

b In some arachnids, in pycnogonids, some small crustaceans, pauropods, and some larval insects (especially aquatic ones) respiratory exchange occurs across the body surface.

Class Merostomata. Movements of the legs maintain a current of water over the gill-books. The rate of ventilation is proportional to oxygen concentration and ceases in anoxic conditions. *Limulus* is an oxygen conformer.

Class Arachnida. In the lung-books (Fig. 22.20A) oxygen diffuses across the walls of the leaves, from the lumina into the blood, which travels to the heart in pulmonary veins. The change in tension of carbon dioxide in the blood stimulates spiracular opening.

Class Crustacea. In aquatic species respiration occurs at the surface of the gills (Fig. 22.20B,C). The class includes oxygen conformers whose uptake increases linearly to the ambient oxygen concentration (e.g.

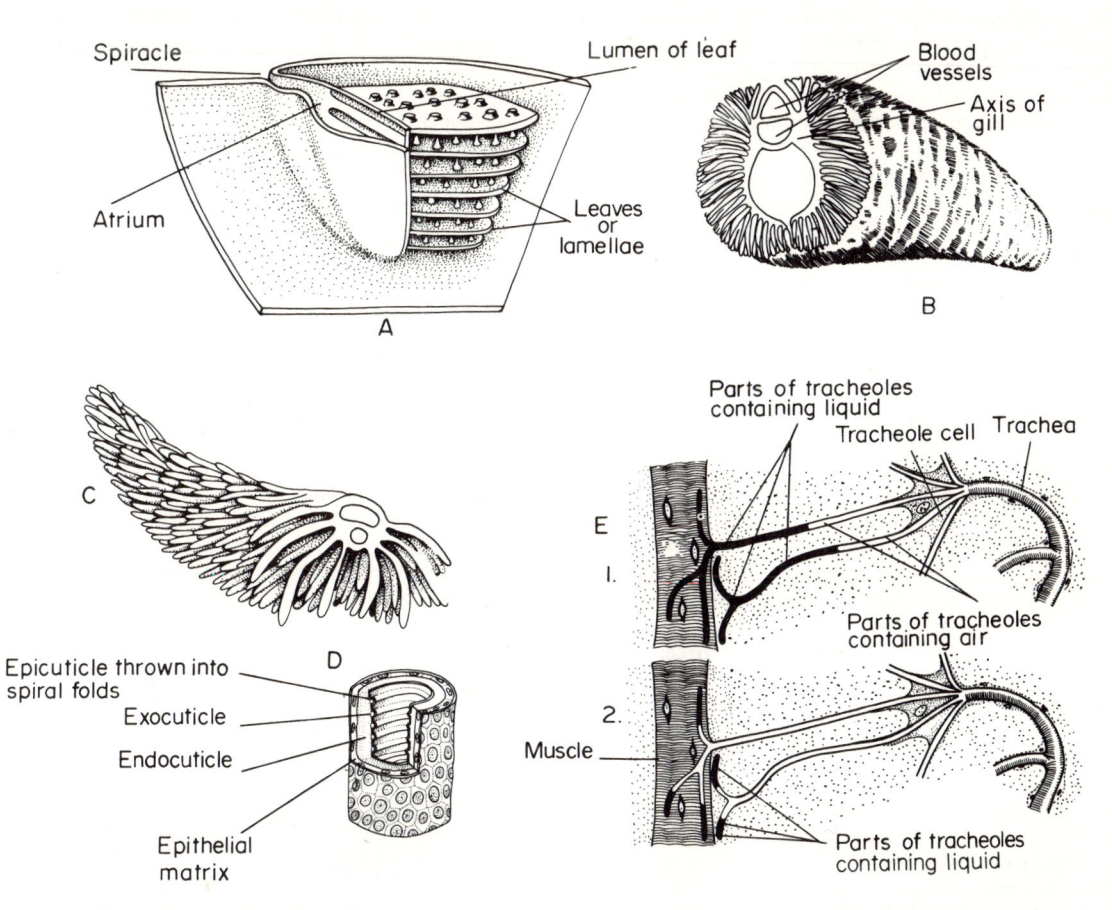

Fig. 22.20 Some respiratory organs of arthropods. A. Lung-book of a spider (after Gerhardt and Kästner). B. Dendrobranchiate gill of a decapod crustacean with 2 series of primary filaments. Each primary filament branches into many smaller filaments. C. Trichobranchiate gill of a decapod crustacean with a series of unbranched filaments arranged around the gill axis (B and C after Meglitsch). D. Structure of an insect trachea, in a region close to the spiracle (after Weber). E. Diagram showing movement of liquid in the tracheoles of an insect. The liquid is absorbed from the tracheoles following metabolic activity in the muscles or other tissue, allowing air to extend towards their extremities and close to the cells requiring oxygen (after Wigglesworth).

Phylum ARTHROPODA

Homarus), and oxygen regulators whose uptake is largely independent of concentration down to some critical pressure (e.g. *Uca*). The rate of water-pumping for respiratory purposes may vary with oxygen or carbon dioxide concentration. Even where special respiratory organs occur, some respiration may occur across the general integument.

Tracheae occur in the classes *Insecta*, *Diplopoda*, *Chilopoda*, and *Symphyla* and in some members of the class *Arachnida* (where tracheae alone may be present, or together with lung-books). The tracheae deliver oxygen directly to the tissues. The tracheae in insects (Fig. 22.20D) terminate in fine, tubular tracheoles (several may be derived from a single tracheal epithelial cell) which end at the cell surface (Fig. 22.20E) or penetrate the cell. Thus oxygen is supplied direct to the cells.

In inactive forms passive diffusion of oxygen through the tracheal system may occur but generally muscular contraction aids the air-flow. Spiracular movements may be controlled by muscles and they may be co-ordinated with abdominal ventilation to drive air through the tracheal system along set paths. The compression of air-sacs (where present) forces stagnant air out through the spiracles, though some loss of carbon dioxide also occurs across the integument.

Circulation/coelom

Arthropods have a much-reduced coelom but possess an extensive haemocoel, derived from the embryonic blastocoel. There is an open circulatory system in which blood, pumped anteriorly by a dorsal, contractile heart, bathes the organs directly. Blood enters the heart through paired openings or ostia. Within the different classes of the Arthropoda the degree of organization of the circulatory system into vessels (arteries, veins, etc.) varies. Much of the blood-flow is through haemocoelic cavities or sinuses, and is more or less directed.

Haemocyanin is the blood pigment in the Merostomata and some arachnids and crustaceans, and haemoglobin in other crustaceans and a few insect species. Pauropodans have no circulatory system.

Class Merostomata. A well-developed circulatory system with arteries and veins supplies all parts of the body. The venous system empties into ventral sinuses, whence blood passes to the gill-books.

Aerated blood flows from the gill-books to the heart in the branchio-pericardial veins, entering through eight pairs of ostia. The pacemaker ganglion initiates heartbeat. The haemocyanin of the blood shows a negative Bohr effect with carbon dioxide enhancing oxygen uptake.

Class Arachnida. The circulatory system is of arteries, veins and sinuses. Blood is pumped from the heart into the arteries and bathes tissues and organs. From the haemocoel blood enters the lamellae of the lung-books, is aerated and flows in the pulmonary veins to the pericardial space. The contraction of cardiac supporting ligaments causes expansion of the heart and blood from the pericardial space enters through ostia.

A lymphatic system is present in some orders.

Class Pycnogonida. The tubular heart with two or three pairs of ostia, lies dorsally above a membrane dividing the body cavity into dorsal and ventral portions. Blood is propelled anteriorly by the heart, circulates in the haemocoelic space, passes backwards in the ventral division, enters the legs and percolates back to the dorsal division through pores in the membrane.

Class Crustacea. The heart extends over a variable number of segments and lies in a pericardial sinus. Arteries, veins, capillaries and sinuses may occur.

Blood is pumped anteriorly from the heart into the arteries (Fig. 22.21) and thence to the sinuses. Returning blood flows from the sinuses into the pericardial sinus and enters the heart through ostia. The heartbeat is neurogenic with ganglion cells in the muscular heart wall initiating the rhythm.

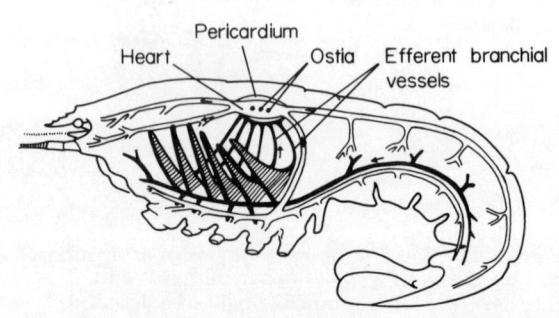

Fig. 22.21 The circulatory system of a lobster, with arteries shown in white, venous channels in black and arrows indicating the direction of blood flow (after Gegenbauer).

Classes Diplopoda, Chilopoda, Symphyla. The diplopods have an elongate heart with some arteries and two pairs of ostia per diplosegment. The chilopod heart is a long tube with ostia in most segments, connected by a pair of aortic arches to a ventral vessel. In symphylans, the heart has ostia in each segment but there is little arterial development.

Class Insecta. The tubular heart, generally with a pair of ostia per segment, is produced anteriorly as the anterior aorta which opens into the head cavity. Other vessels may be present. When ostia are lacking blood returns to the heart through fine, open-ended vessels. A diaphragm (more or less complete) just ventral to the heart separates the body cavity into

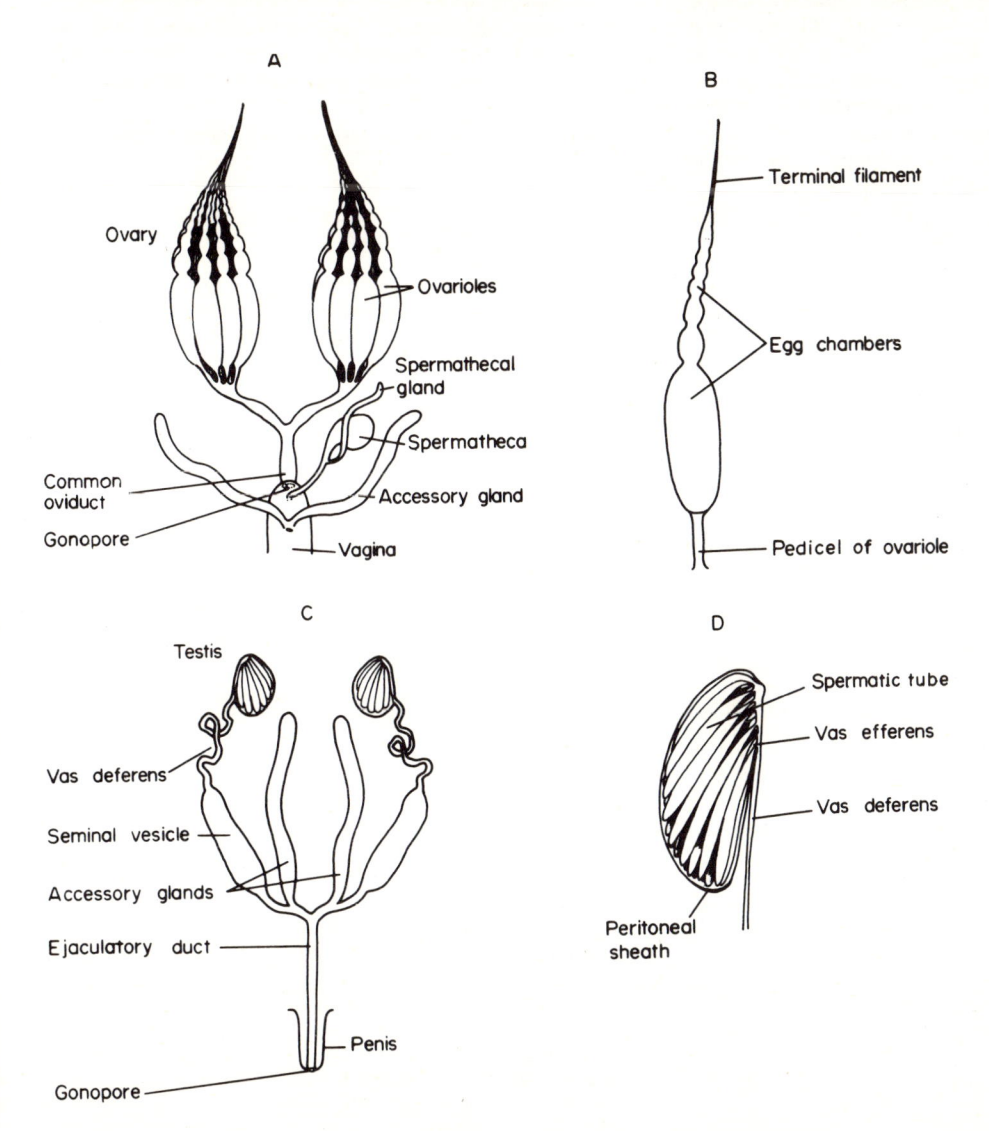

Fig. 22.22 Reproductive organs of insects. A. Diagram of the female organs. B. A single ovariole, enlarged. C. Diagram of the male organs. D. A single testis, enlarged (A—D after Snodgrass).

Phylum ARTHROPODA 127

pericardial and perivisceral sinuses. A diaphragm above the ventral nerve cord may also be present.

Blood, pumped anteriorly, passes from the anterior aorta into the head and body sinuses. Movement of blood in the haemocoel is assisted by body movements, and muscular contractions of body wall, gut and the muscular diaphragms. The muscles associated with the pericardial diaphragm expand the heart after contraction. The heartbeat is neurogenic.

In some insects accessory 'hearts' may occur which aid circulation of blood through legs and wings.

In insects the blood plays a minor role in the distribution of oxygen to the tissues. Various types of haemocyte occur in the blood.

Reproduction

a In the Arthropoda the sexes are typically separate (Table 22.5) but a few crustacean species are hermaphroditic. Reproduction is typically sexual but parthenogenesis occurs in some insects and crustaceans. Sexual dimorphism is common. There is internal fertilization in terrestrial forms but external fertilization occurs in some aquatic species. The male may possess appendages modified for use during copulation, e.g. for clasping, transferring sperm etc. The ova are typically centrolecithal (central yolk) and undergo superficial cleavage.

Table 22.5 Arthropod reproductive organs.

Class	Ovary	Testis	Ducts and genital openings	Accessory structures which may be present
Merostomata	Tubular network	Tubular network	Short ducts lead to paired openings on the inside of the genital operculum (segment 8)	
Arachnida	Single or paired, form variable	Single or paired, form variable	Paired ducts unite to give a median uterus or median male organ with a single ventral opening protected by epigynum, usually on 8th body segment; penis present in groups with direct internal fertilization	Spermathecae associated with vagina; ♂ accessory glands involved in spermatophore formation
Pycnogonida	Single, U-shaped, branches in legs	Single, U-shaped, branches in legs	Genital openings situated at bases of legs; number, and specific legs vary with species and sex	
Crustacea	Paired, elongate, dorsal	Paired, elongate, dorsal	Paired simple ducts open at limb bases or ventrally on a sternite, segment varies with species	
Diplopoda	Single, tubular, ventral	Paired, tubular, ventral	♀ with median oviduct and uterus bifurcating to 2 vulvae opening ventrally near coxae on 3rd segment; ♂ sperm ducts open by penis (paired or single) ventrally on 3rd segment	Seminal receptacles associated with vulvae

(Continued on page 129)

Table 22.5 (*continued from page 128*)

Class	Ovary	Testis	Ducts and genital openings	Accessory structures which may be present
Pauropoda	Paired, ventral	Paired, dorsal	♀, united oviducts open into depression between legs of 3rd segment; ♂ sperm ducts open by paired penes between coxae on 3rd segment	Seminal receptacle opens into depression
Chilopoda	Single, tubular, dorsal	1—24 dorsal testes	♀, single duct opens on apodous genital segment; ♂, paired sperm ducts open through a median pore at tip of penis, on ventral side of genital segment	
Symphyla	Paired, lateral	Paired, lateral	Gonopores open ventrally on 4th trunk segment	
Insecta	Paired, consist of grouped ovarioles (Fig. 22.22A)	Paired (Fig. 22.22B)	♀, paired oviducts unite, common median duct leads to vagina opening on 8th or 9th abdominal segment; ♂, each vas deferens leads to a wider seminal vesicle and into a common median ejaculatory duct opening by a penis, ventrally, on abdominal segment 9	(i) Spermathecae associated with vagina (ii) Accessory glands of vagina involved in shell formation (iii) Sperm storage-chamber associated with ejaculatory duct (iv) Accessory glands opening into seminal vesicles produce fluid medium for sperm

b

Class Merostomata. Maturity is reached in the third year and mating occurs, the male clasping the female with the modified first prosomal legs. The eggs are laid in sand, in the inter-tidal zone, and fertilization is external. Hatching takes several months.

Class Arachnida. Internal fertilization occurs in some species but sperm transfer is characteristically indirect, a spermatophore being commonly employed. Pre-copulatory 'courtship' behaviour is often complex.

Where sperm transfer is accomplished without spermatophore formation, the male has the tarsus of the pedipalp modified for sperm storage and transfer (Fig. 22.23). Sperm are ejected onto a special web pad, and pumped into a palpal receptaculum seminis whose spine-like duct or embolus is inserted into the female opening.

Where a spermatophore is employed the female may be drawn over the deposited spermatophore, or spermatophore transfer may be accomplished by the chelicerae of the male or a modified region of the 3rd pair of legs.

The fertilized eggs are protected in a burrow, nest, cocoon etc. and may be deserted, guarded or carried by the female. Scorpions are viviparous.

Class Pycnogonida. Fertilization is external. The eggs are picked up by the male and attached to the ovigerous legs by a secretion from cement glands occurring in the 4th joint of some legs. Here the eggs are brooded.

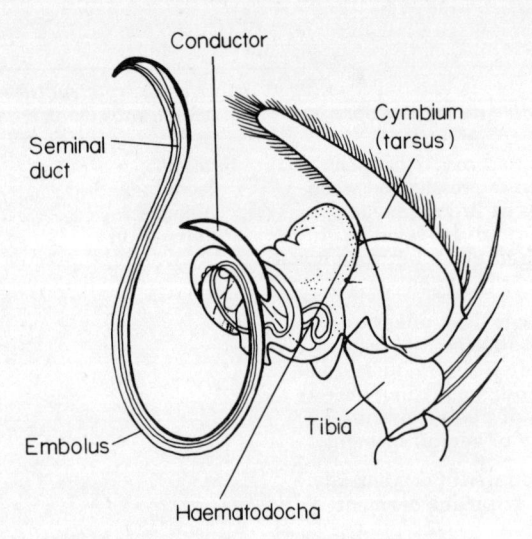

Fig. 22.23 Distal segments of the pedipalp of a male spider *Agelena naevia* (Araneae, Arachnida), showing the palpal intromittent organ (after Petrunkevitch).

Class Crustacea. Mating behaviour may be complex with pheromones present in some groups, and mating often occurs only at the time of the female moult. Copulation typically occurs, the male generally using appendages to grasp the female or to transfer sperm or spermatophores. Some species release fertilized eggs into water but often the eggs are brooded, e.g. in a special pouch (Cladocera), attached to external hairs (Decapoda), in an egg sac (Copepoda) or mantle cavity (Cirripedia).

Class Diplopoda. There is 'courtship' behaviour in some species. In most diplopods the 7th diplosegment bears a pair of modified legs, the gonopods, which take up sperm from the penes and act as intromittent organs which transfer sperm to the vulvae. Fertilization occurs when the eggs are laid. The eggs are laid in soil or into a nest and may be guarded by the female.

Class Pauropoda. Little is known of mating behaviour. The eggs are laid in humus.

Class Chilopoda. 'Courtship' behaviour occurs in some species. Sperm transfer is indirect. The male deposits a spermatophore on a special web and the female walks over it, picking up the spermatophore in her genital aperture. Manipulation of the spermatophore is aided by the gonopods on the genital segment. The eggs are laid in decaying wood or soil and in some species are guarded by the female.

Class Symphyla. Little is known of mating behaviour and fertilization. *Scutigerella* employs a spermatophore.

Class Insecta. 'Courtship' behaviour is often complex, with sexual attractants or pheromones being produced in some species. Typically sperm transfer is direct, with the everted penis of the male injecting sperm into the female vagina. Modified appendages on the genital segments may function as claspers in the male and as an ovipositor in the female. Since many insects mate only once (often before oviposition) sperm are stored in the spermathecae and released as required. Ova are fertilized in the vagina, prior to laying, and the site of oviposition is variable.

Economic importance

The arthropods include many species of great economic importance to mankind, some beneficial, some detrimental.

Certain groups (particularly the decapod crustaceans) are an important source of food for man. Insects are essential, as pollinators, for the perpetuation of many flowering plants, including important food crops. Predatory insects which feed on other species may play a role in the control of certain insect pests.

Amongst the insects are included parasites of man, domestic animals, ornamental plants and food crops. Insect vectors are responsible for transmitting disease in man, animals, and plants. Other species feed on standing crops (e.g. locusts), stored food (e.g. cockroaches) or attack structural or other timber.

A major problem at the present time is the unsophisticated use of insecticides which not only poison the environment but fail to discriminate between beneficial and harmful species. Given the high reproductive rates of insects, application of pesticides rapidly leads to the production of resistant forms, thus necessitating even higher doses of pesticide. The futility of this method of pest control should be obvious.

References

Bullock T.H. & Horridge G.A. 1965. *Structure and Function in the Nervous Systems of Invertebrates*, Vol. 2. W.H. Freeman and Co. San Francisco and London.

Clarke K.U. 1973. *The Biology of the Arthropoda*. Edward Arnold, London.

Rockstein M. (Editor) 1964. *The Physiology of Insecta*, 3 vols. Academic Press, New York and London.

Waterman T.H. (Editor) 1960. *The Physiology of Crustacea*, 2 vols. Academic Press, New York and London.

23 Phylum TARDIGRADA

Very small animals, up to 1 mm in length. Some 200 species exist.

Characteristics

1 Majority inhabit fresh water, occurring on mosses, lichen and other small plants. A few are marine.
2 Some show signs of segmentation.
3 Each segment has a pair of claws.
4 Mouth has a sucking proboscis with stylets.
5 Possess a non-chitinous, cuticular exoskeleton.
6 Undergo periodic moults.
7 Total number of cells in the body is probably fixed.

Larval form

The tardigrade juvenile is small with a head and 4 segments; 4 pairs of legs are developed before hatching (Fig. 23.1).

Metamorphosis

Tardigrades hatch as juveniles with a given number of cells. Growth occurs by increasing size of cells.

Adult body form e.g. *Macrobiotus, Echiniscus*.

These are short-bodied, fat animals. There are four body segments, each bearing a pair of short legs, and the head is small. The legs end in claws. The body is covered by a cuticle, smooth or bristly, which is totally shed from time to time. The cuticle contains no chitin. Some signs of metamerism may be noted, the cuticle being arranged in plates (*Echiniscus*), muscles demonstrating repetition, and the nervous system having four ganglia (Fig. 23.2).

Feeding

a Tardigrades are mainly herbivorous. They devour the liquid content of plant cells. Some have been observed to feed on small invertebrates.

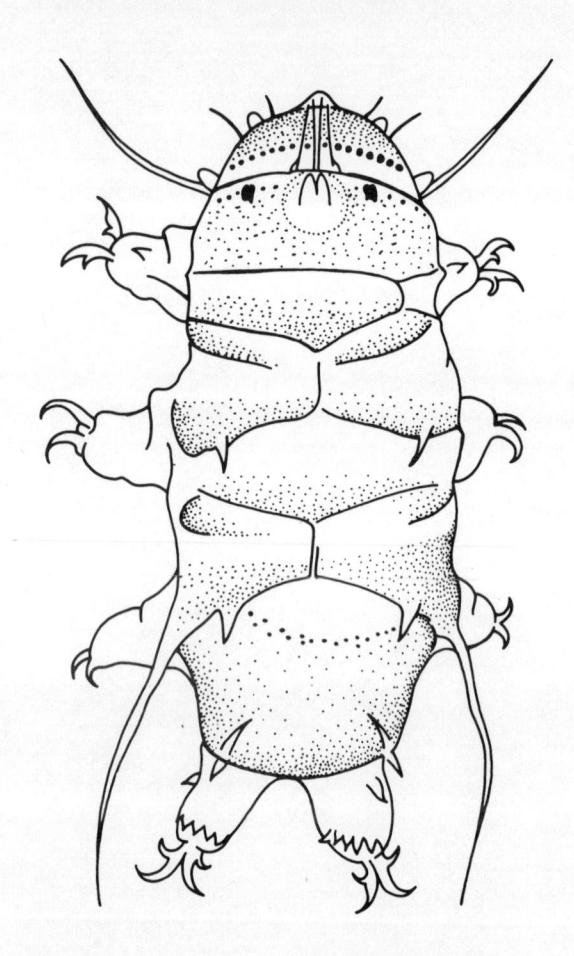

Fig. 23.1 Larva of *Echiniscus granulatus* (after Cuénot).

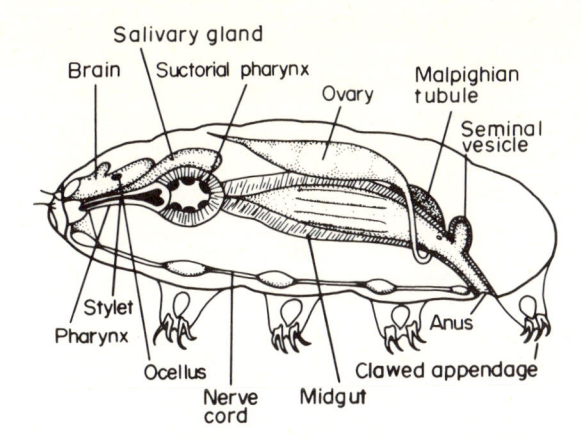

Fig. 23.2 Schematic diagram of tardigrade anatomy (after Meglitsch).

b The gut opens at an anterior mouth, leading into a buccal cavity covered with stylets, these being secreted by a special gland prior to each moult. The anterior gut is adapted for (i) piercing the cellulose wall of plant cells, by means of two sharp stylets on the buccal tube, and (ii) sucking out the contents, using a muscular suctorial pharyngeal bulb. There is a capacious midgut, and a short rectum opening via a terminal anus. There are three glandular sacs joining the hindgut, which may be excretory.

Osmoregulation/excretion

a The three glandular areas, known as Malpighian tubules, joining the gut at the mid-hindgut junction, may be excretory organs. The gut itself may be excretory since faeces and other materials are voided with the cuticle at moult.

b Regulatory physiology is unknown. Survival under extreme environmental conditions is remarkable. Desiccation, immersion in liquid helium, brine etc. are not necessarily lethal. During diapause tardigrades lose much of their volume and, presumably, water content

Movement

a Movements are limited, there being only slow crawling.

b Attachment is by the small claws on the legs.

c Muscles occur in bands rather than layers. Muscles are single, smooth muscle fibres. They arise from the walls of the coelomic pouches that disintegrate early in development.

Co-ordination

a i For such unpretentious animals the nervous system is well organized. There is an anterior brain with large lateral lobes joined by connectives around the anterior gut to a sub-pharyngeal ganglion. From this extends a double nerve cord, with four ganglia along its length. Several pairs of lateral nerves arise at each ganglion. There are eye spots, and bristles and spines may be sense organs. Most neurones seem to be unipolar.

ii The physiology is not known.

b Hormones are not described but the phenomenon of moulting seems presumptive evidence for complex endocrine interactions.

Respiration

a No respiratory organs have been described.

b The physiology is not known but the tolerance of extreme conditions, and long periods of diapause (anabiosis) argue for most interesting adaptations of metabolism.

Circulation/coelom

a There is no circulatory system. The body cavity is a fluid-filled pseudocoel. The coelomic pouches appear during the embryonic stage but only one pair is retained (as the gonad) in the adult.

b Fluid propulsion is due to movement of body wall and gut muscles.

c There are no respiratory pigments.

Reproduction

a The sexes are separate. Females are usually more numerous than males and in some species males are not known.

One gonad is found. The testis has two ducts that open at a single gonopore in front of the anus. The ovary has one oviduct opening above the anus, or into the rectum (in which case a seminal receptacle also occurs).

b Copulation takes place, fertilization is internal, and may occur at a moult. Eggs may be thin-shelled and develop immediately, or thick-shelled for resistance to poor external conditions. Some tardigrades may be parthenogenetic. Development is direct. Coelomic pouches arise enterocoelically.

Anabiosis

The ability of tardigrades to withstand astonishing environmental adversity is almost legendary, and equally almost totally mysterious. No satisfactory investigations have been carried out on the manner in which the anabiotic longevity is attained.

Reference

Grassé P. P. (Editor) 1949. *Traité de Zoologie*, Vol. VI. Masson et Cie, Paris.

24 Phylum PENTASTOMIDA

All members of this phylum are small (up to about 10 cm), worm-like animals. There are about 60 species.

Characteristics

1 All are parasites of carnivorous vertebrates.
2 There are five short bumps anteriorly, one bearing the mouth, the others claws.
3 There is a cuticle.
4 The cuticle is periodically moulted.
5 Life-cycle shows interpolation of an intermediate host (herbivore).

Larval form

The pentastomid larva has two pairs of legs, bearing claws (Fig. 24.1). The eggs in which the larvae develop lie in vegetation.

Metamorphosis

When the vegetation is grazed by a small, vertebrate herbivore (e.g. a rabbit or fish) the larvae are released in the gut. They migrate through the gut wall, circulate in the blood and eventually undergo a series of moults whilst encysted in the liver. At this stage the juvenile is like the adult but covered with many small bristles. If now the herbivore is eaten by a carnivore (dog, crocodile etc.) the larvae are released, and attain a final position in the nasal cavity or lungs.

Adult body form e.g. *Linguatula, Porocephalus.*

Pentastomids (also known as linguatulids) are elongate animals, some like flukes, some with external annulations, while others have no segmental appearance. The two pairs of limbs prominent in the larva are found only in

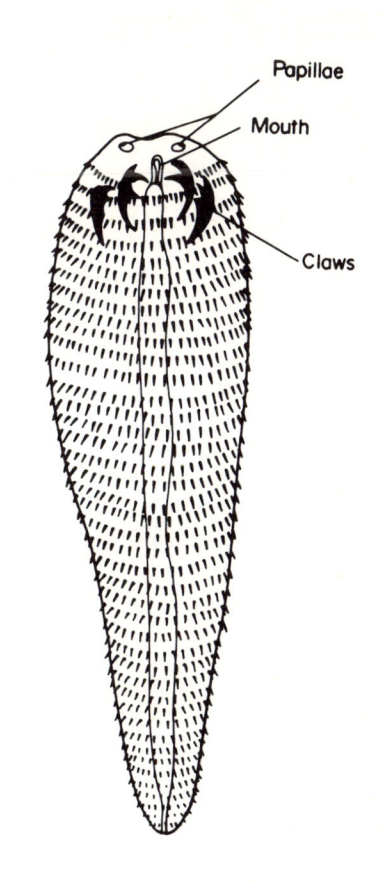

Fig. 24.1 Larva of *Linguatula* (after Korschelt and Heider).

Cephalobaena adults (Fig. 24.2), the other genera showing progressive loss until only the claws remain. There is a thick, chitinous cuticle covering the body, and beneath it lie 2 layers (circular and longitudinal) of striated muscle in the body wall. There are no circulatory, respiratory or excretory systems. Sexual organs are, however, well developed (Fig. 24.3).

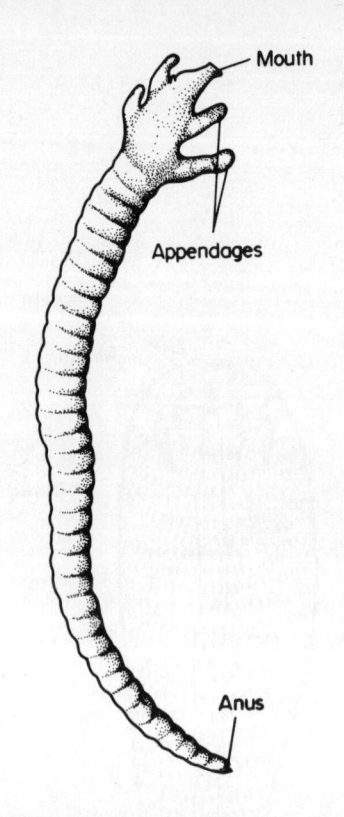

Fig. 24.2 *Cephalobaena tetrapoda* (from the lungs of snakes), retaining the four limbs (after Heymons).

Feeding

a Pentastomids are permanently attached to the tissues of the host, in the aerial passageways and lungs. They suck blood from the circulatory system.

b The anterior gut is modified for a suctorial function, the remainder is a straight tube.

Osmoregulation/excretion

a No excretory organs exist.

b Excretion probably takes place via the gut, osmoregulation in a relatively stable environment may be minimal.

Movement

a Little movement apart from local body wall flexing is required, the animal remaining permanently in one situation. The larva can bore through tissues of the intermediate host and eventually in the definitive host, climbs from the oesophagus to the nasal cavity.

b Movement and attachment utilize the small claws of the anterior end.

c Muscle physiology is not known but all muscle fibres are striated. Fibre structure is simpler than for many free-living invertebrates.

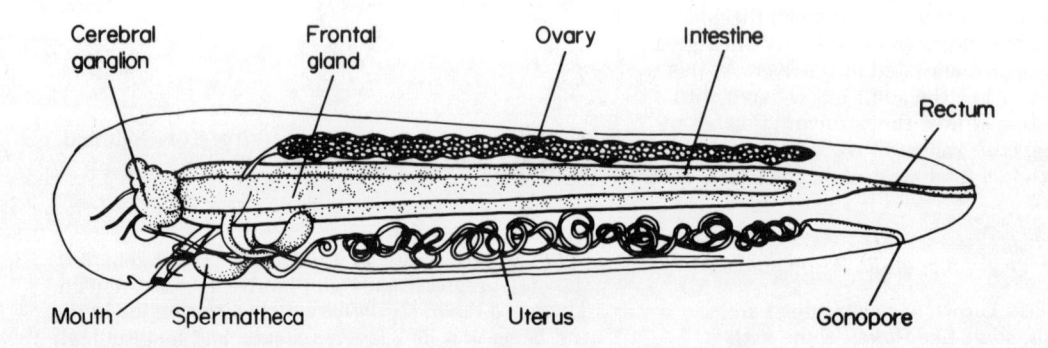

Fig. 24.3 Organs of *Waddycephalus* ♀ from lateral aspect (after Spencer).

Phylum PENTASTOMIDA

Co-ordination

a *i* The nervous system is composed of an anterior circumoesophageal ring, and 3 metameric ganglia along a ventral nerve cord. Sense organs are borne on papillae on the body surface.

ii No details of nervous function are available.

b The presence of moulting hormone, in early stages, is implied in the frequency of larval moults.

Respiration

a There are no respiratory organs.

b Gaseous exchange must be a function of the general body surface.

Circulation/coelom

No circulatory system is present. Pentastomids are coelomate.

Reproduction

a The sexes are separate.

b Fertilization is internal, sperm are stored in the seminal receptacles until required. The laid eggs are embryonate (already developing) and may be voided from the host respiratory system by sneezing (*Linguatula*) or by passage through the gut and out with the faeces. The egg then must be taken up by the intermediate host.

Reference

Grassé P.P. (Editor) 1949. *Traité de Zoologie*, Vol. VI. Masson et Cie, Paris.

25 Phylum MOLLUSCA

Molluscs account for some 110,000 species. Large specimens may be measured in metres (lamellibranchs, cephalopods) whilst the smallest are microscopic in size.

Characteristics

1 Molluscs inhabit marine, fresh-water and terrestrial habitats.
2 The typical larva is of a modified trochophore type.
3 Typically there is a head, a muscular foot and a visceral hump.
4 The hump is covered by a mantle which often secretes a shell (composed of an organic matrix which is calcified) and encloses a cavity into which open the anus and kidneys.
5 The alimentary canal has a muscular buccal mass, radula, salivary glands and a stomach into which opens the digestive gland. There may be a crystalline style.
6 The nervous system is a circumoesophageal ring, (which may be condensed into a central mass of nervous tissue, containing cerebral and pleural ganglia), pedal cords, and visceral loops.
7 There are ctenidia, originally used as respiratory organs, but in some groups adapted as feeding devices.
8 Molluscs are coelomate but the coelom is often greatly reduced in extent. It is always present as the pericardium, the cavity of the kidneys (linked to the pericardium) and the gonadal cavity.
9 There is a blood system, with a propulsive heart (median ventricle and 2 lateral auricles reduced in some species to one), and an arterial and venous system that often opens into an extensive haemocoel. The respiratory pigment of the blood is haemocyanin (Cu^{++}-containing).

Larval form

The molluscan larva is readily compared with the trochophore of the annelids. The gastropod veliger larva is most closely similar with a velum composed of a tissue flap bordered by large motile cilia that provide locomotory power (Fig. 25.1A). The major portion of the body develops behind this velum and forms the visceral hump, which shows no sign of segmentation.

There are variants upon this theme; the chitons and scaphopods have large, yolky eggs that form embryos with a prominent prototroch ring of cilia; bivalves form a glochidium that is a small version of the adult, but with a ciliated velum and showing much active swimming; cephalopods demonstrate direct development to a juvenile within the egg sac.

Metamorphosis

Cephalopods show no metamorphosis. Eggs develop directly into juveniles, hatching as immature

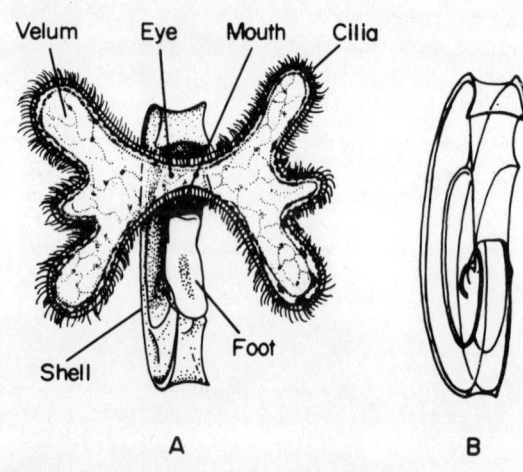

Fig. 25.1 A. Echinospira veliger of *Lamellaria perspicua* seen from the front in the swimming mode to show extent of velum. B. The transparent vacated shell of *L.perspicua* (A and B after Fretter and Graham).

individuals. They may all (even octopods) spend a planktonic period immediately after release before becoming mature. Lamellibranchs also show very little change in morphology from larva to adult. The larva is bivalved at an early stage but possesses a velum for swimming. This is lost at metamorphosis when the larva settles and uses only the foot for propulsion.

The greatest metamorphic changes are exhibited by gastropods. The early veliger quickly develops spiral coiling of the viscera, followed by torsion during which the visceral mass is rotated above the foot. The mantle cavity moves from the primitive posterior position to an anterior position. The initial stage of torsion may be a rapid event, muscular in origin and taking only few hours, completion being due to slow growth. In other cases it is a growth phenomenon only and may take several days. It may take place during a free larval life, usually accompanied by a change in behaviour, or during encapsulated development.

After torsion the mantle cavity and its contents face forward, the auricle of the heart lies in front of the ventricle (Fig. 25.2D, E) amd the connective nerves are crossed over one another. In the opisthobranchs these changes show regression.

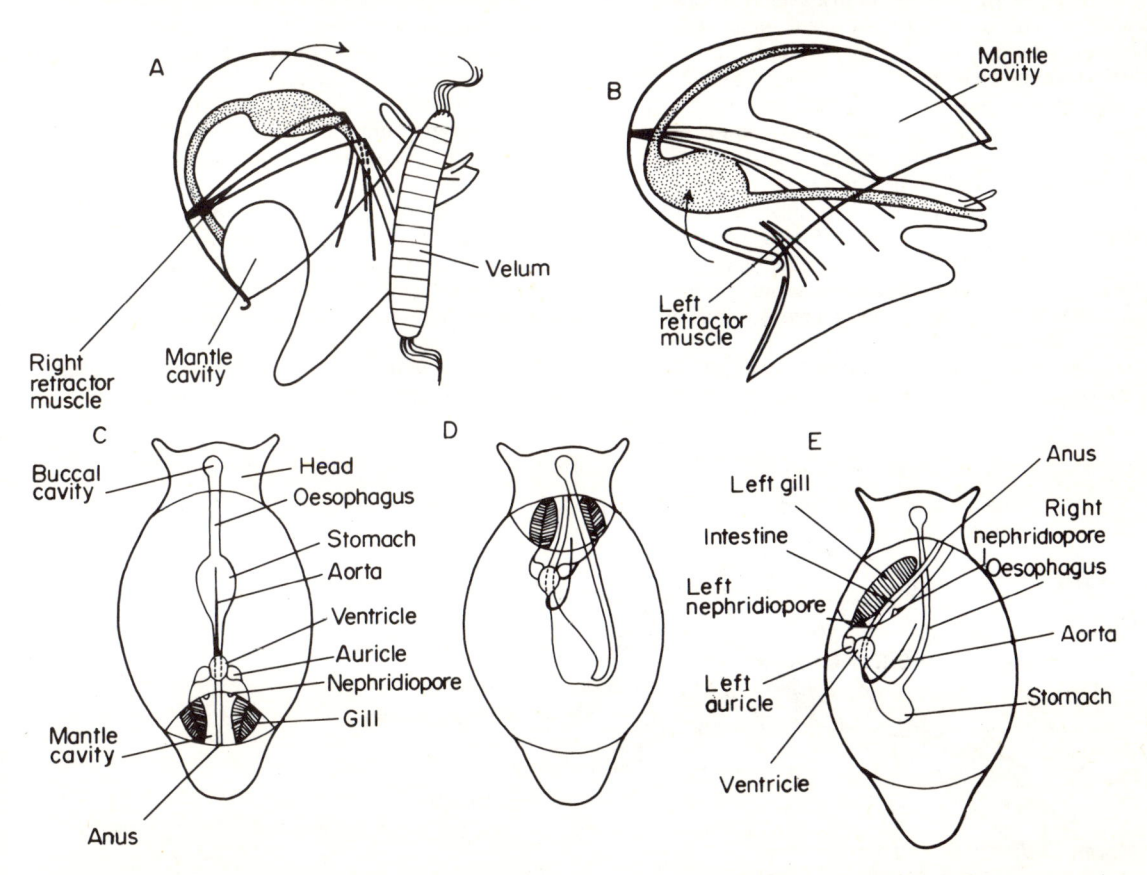

Fig. 25.2 The process of gastropod torsion and its consequences. A. The early veliger, with posteriorly-placed mantle cavity and locomotory velum. The asymmetrical retractor muscle is shown. B. Consequent upon contraction of shell muscles (and further slow growth) the mantle cavity lies anteriorly (A and B after Crofts). C. Indicates ancestral condition before torsion. D. After torsion (ancestral). E. Situation in a prosobranch gastropod. Note reduction in ctenidium, and associated organs (C–E after Graham).

Phylum MOLLUSCA

Adult body form

It is often stated that despite the variety of bodily types shown amongst the Mollusca all can be derived from a single ancestral 'type'. This theoretical animal incorporates all the diagnostic characters of the group arranged in what purports to be a primitive manner. Such an example is shown in Fig. 25.3. From this form the extant classes have departed to greater or lesser extent.

1 *Class* MONOPLACOPHORA e.g. *Neopilina*.
These were long known only as fossil forms, but in 1952 living representatives were discovered by the 'Galathea' expedition in the deep-sea trenches off the shores of Costa Rica. Monoplacophora are bilaterally symmetrical molluscs, having a central foot and a median posterior anus; the mantle is covered by a one-piece shell which in the embryo shows signs of dextral coiling; within a shallow mantle cavity 5 pairs of branchiae are found. This multiplication of organ systems is typical also of the vascular (2 pairs of auricles), renal (6 pairs of kidneys), reproductive (2 pairs of gonads) and muscular (8 pairs of shell muscles) systems (Fig. 25.4B). The nervous system resembles that of chitons and is simple.

The repetition of organs may be an indication of metamerism (as in annelids) but if so is not in any way demonstrated externally, nor in the anatomy of the nervous system, nor in a regular fashion (all systems having different numbers of pairs). The phenomenon may therefore represent multiplication as found in other archaic molluscs (*Nautilus*, chitons) although the functional significance of these repeated structures is not clear.

2 *Class* AMPHINEURA e.g. *Lepidochiton*, *Mopalia*.
Possess a head of primitive type but without tentacles and eyes. The foot is flattened and moves by waves of contraction passing forward. The Amphineura show bilateral symmetry. The mouth and anus lie at opposite ends of the body, the visceral mass between being covered by a pallium in which lie chemoreceptors, and which is loaded with spicules that may unite to form valves (Fig. 25.5A) (8 in all in Polyplacophora, chitons). The nervous system has no ganglia.

The Aplacophora (Solenogastres) e.g. *Proneomenia*, *Cryptoplax*, are distinct by virtue of their embryology, and in general are vermiform animals (non-segmented) that exhibit primitive and specialized characters. The mantle cavity is a shallow depression posteriorly; there is a thick cuticle, but no shell; there may be a radula or a suctorial bulb. All are specialized feeders, and some are abyssal.

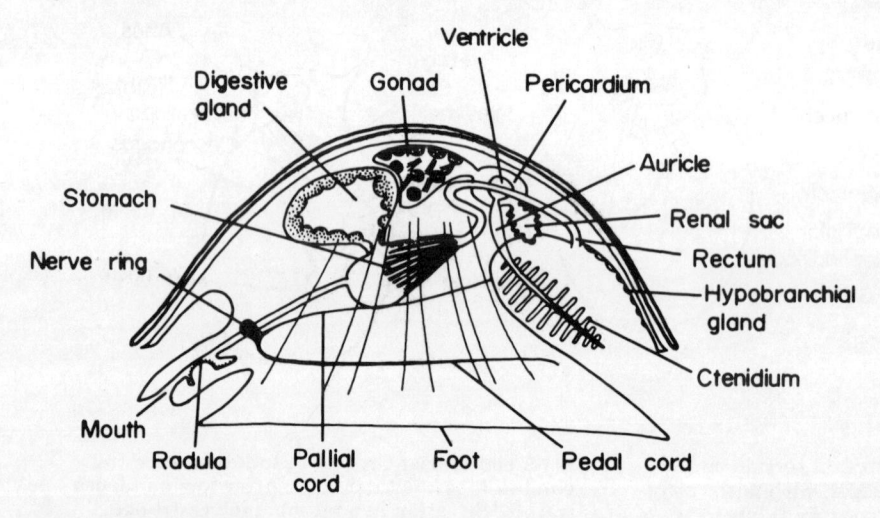

Fig. 25.3 Basic molluscan plan from which the existing forms may be derived. It shows all the major features in a primitive condition, each of which has been modified in characteristic ways in the present-day classes (after Morton & Yonge).

Phylum MOLLUSCA

separate sexes, often with an operculum. The primitive forms are the Archaeogastropoda (*Haliotis, Patella*) in which some of the original bilateral symmetry is retained (e.g. 2 ctenidia, 2 auricles). More advanced genera are members of the Mesogastropoda (*Viviparus, Littorina*) with asymmetry of the palliopericardial complex (reduction to 1 monopectinate ctenidium and 1 auricle), concentrated nervous system, free-swimming veliger larva and a siphonate shell in some. The most advanced group is the Neogastropoda (*Buccinum, Nucella*), possessing a highly concentrated nervous system, a siphonate shell and an eversible proboscis (all are carnivorous or scavengers). Development is intracapsular.

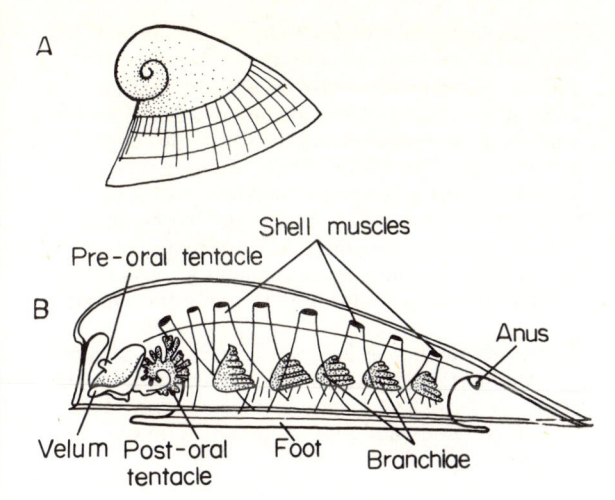

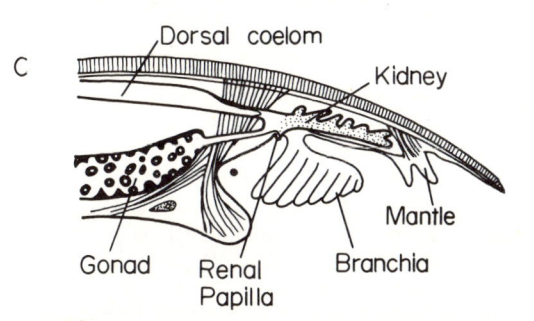

Fig. 25.4 *Neopilina*. A. Embryonic shell. B. Schematic lateral view. C. Transverse section of right half of pallial cavity (A—C after Lemche & Wingstrand).

3 *Class* GASTROPODA   e.g. *Buccinum, Patella, Aeolidia, Helix.*
Head as in primitive form, with tentacles and eyes. The foot is flattened and creeping (Fig. 25.5B) except in pteropods where it is winglike and expanded for floating and swimming. The body is asymmetrical, showing torsion and coiling of the visceral hump (indicated by the pallium and shell). This has resulted in reduction and asymmetry of the vascular system.

The class is usually divided into 3 sub-classes, as below.

PROSOBRANCHIA   generally aquatic, with pronounced torsion, crossed visceral loop in the nervous system, mantle cavity opening anteriorly,

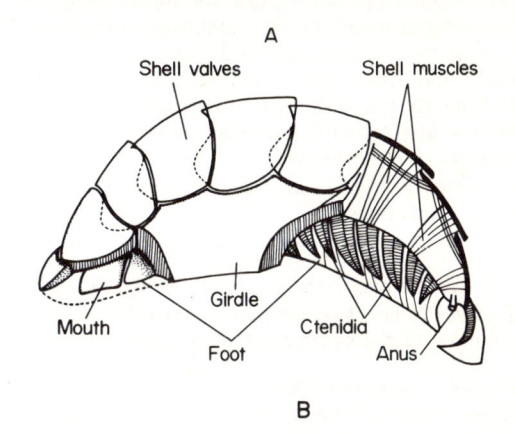

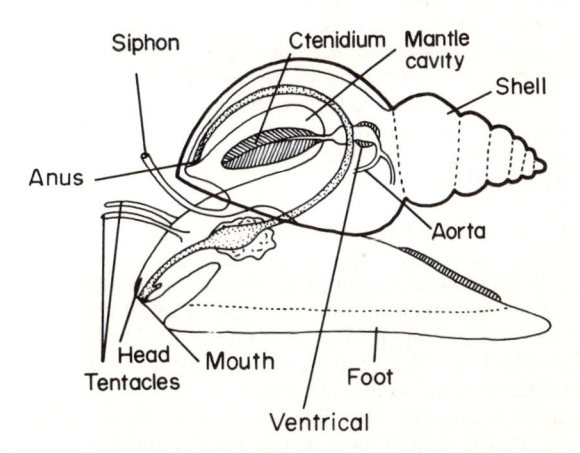

Fig. 25.5 Characteristic features of A. an amphineuran (after Morton & Yonge), B. a gastropod (after Meglitsch).

Phylum MOLLUSCA   141

OPISTHOBRANCHIA   e.g. *Eolis, Doris*. Opisthobranchs are marine and hermaphrodite. The shell, massive in prosobranchs, is much reduced or even lost. Detorsion may take place, and the visceral loop of the nervous system is uncrossed. The animals may exhibit external bilateral symmetry.

PULMONATA   This group inhabits fresh-water and terrestrial habitats. They are hermaphrodite, exhibit torsion and many have a heavy shell, but bear no operculum. The ctenidium is lost, the mantle cavity being vascularized as a lung which opens to the exterior via a small, contractile, pallial aperture, the pneumostome. The nervous system is symmetrical due to central concentration. The group includes the snails and slugs (which lose the shell) in the order Stylommatophora, in which the eyes are at the tip of the posterior, retractile tentacles, e.g. *Helix, Arion*. Fresh-water snails may be in the order Basommatophora, with eyes at the base of the posterior, non-retractile tentacles. Secondary gills may develop. e.g. *Limnaea*.

Two features are of major importance in gastropod organization.

1   *Spiral coiling* of the visceral mass, due to great growth of the digestive glands and looping of the alimentary canal. It bulges into the mantle cavity and one set of pallial organs may be reduced or lost.

2   *Torsion*, the process by which the visceropallial complex swings through 180° to move from the primitive posterior position to the anterior position it holds in many gastropods. The effects are that the ctenidia face forward, the auricles lie in front of the ventricle; the mantle cavity opens behind the head; the connective nerves form a figure 8 with one branch above and one below the alimentary canal, and the body of the animal can now retract into the mantle cavity.

4   *Class* SCAPHOPODA   e.g. *Dentalium*.
The so-called elephant's tusk shells, these animals exhibit bilateral symmetry and possess a tubular shell which opens at both ends (Fig. 25.6A). The head and foot are a functional unit with a prehensile proboscis used for exploration and digging in the substrate. Sense organs are much reduced. There is a radula in a buccal mass, but there are no ctenidia in the mantle cavity (there being instead a ciliated, respiratory epithelial area). There is no pericardium, heart or blood vessels. The larva is of trochophore type.

5   *Class* LAMELLIBRANCHIA   e.g. *Pecten, Nucula*.
The head is reduced or absent with the mouth located beneath the anterior shell muscle. Sense organs are located on labial palps, siphons and mantle edge. There is no radula or buccal mass and the class is one of suspension-feeders. The foot is flattened laterally, being rarely used in creeping, mostly for burrowing instead. It is protruded by blood pressure. Many lamellibranchs are sedentary animals, adhering by the secretion of a foot gland (byssus).

The visceral mass is centrally placed and extends into the foot, with no coiling or torsion. The gonad lies in the foot. The mantle forms two lobes that enclose the body and secretes a shell composed of two valves joined by a dorsal ligament, which constitutes a hinge (Fig. 25.6B). The ctenidia are large, greatly ciliated and involved in feeding processes.

There are three sub-classes, as below.

PROTOBRANCHIA   with flat ctenidia, non-reflected filaments, e.g. *Nucula, Yoldia*.

LAMELLIBRANCHIA   with large ctenidia of long, reflected filaments that may be joined by ciliary junctions (filibranchs) or tissue junctions (eulamellibranchs), e.g. *Mytilus, Ensis, Pholas*.

SEPTIBRANCHIA   with equal adductor muscles, mantle edges not fused and gills changed to a muscular septum that pumps water through the mantle cavity, e.g. *Cuspidaria*.

6   *Class* CEPHALOPODA   e.g. *Alloteuthis, Eledone, Argonauta*.
The head is well developed and integral with the foot. Sense organs are numerous and large. There is a large buccal mass and powerful radula. The foot, comprising tentacles and siphon, is associated with the head, and is used for walking (in octopods) or prey capture (decapods and octopods). The visceral mass is bilaterally symmetrical, there being no coiling, and the enveloping mantle is muscular and pumps water in and out of the mantle cavity. There is a chambered shell, which is much reduced in some forms, that functions in buoyancy control (Fig. 25.7). These are active animals, with a very concentrated and centralized nervous system.

Some living forms are found amongst the Nautiloidea (*Nautilus*) which have an external,

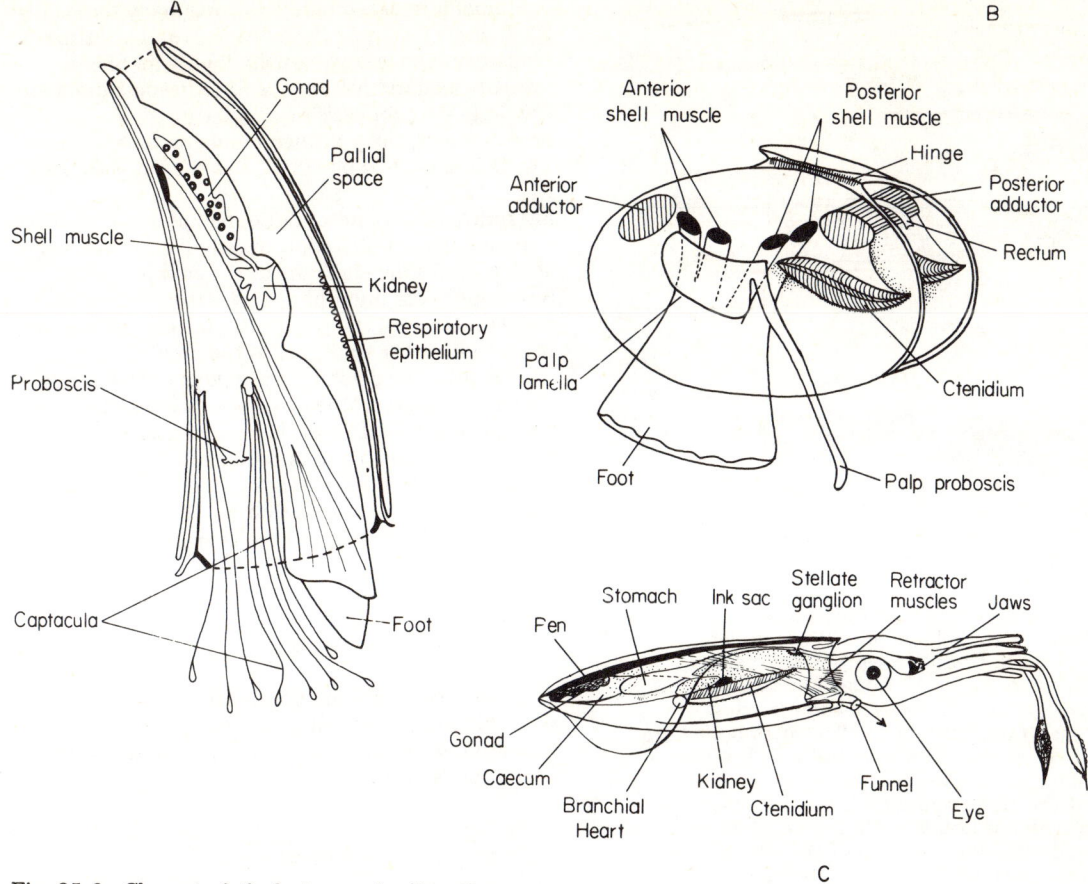

Fig. 25.6 Characteristic features of A. a scaphopod. B. a lamellibranchiate (primitive). C. a coleoid cephalopod (A—C after Morton & Yonge).

chambered shell, numerous unsuckered and retractile tentacles, a funnel formed of two separate folds and are tetrabranchiate with 2 pairs of ctenidia and kidneys. The vast majority of living cephalopods are members of the Coleoidea (Fig. 25.6C), e.g. *Octopus*, *Loligo*. These are dibranchiate (1 pair of kidneys and ctenidia), have an internal shell, 8 or 10 suckered tentacles (2 of which are always long in 10-armed forms), a funnel which is a closed tube, eyes with lenses and an ink sac. Mantle movements are co-ordinated by a system of giant fibres.

The skeleton of all molluscs is of two kinds (some may have extra specialized elements as well):

1 **hydrostatic**, the blood filling large haemocoelic spaces in the body and the varying blood pressure being responsible for the alterations in dimensions of the organs, especially the foot;

2 **the calcareous shell**, formed by deposition of inorganic salts on an organic secreted matrix (conchiolin).

Feeding

a Monoplacophora, most amphineurans and some gastropods are browsers and grazers on small

Phylum MOLLUSCA 143

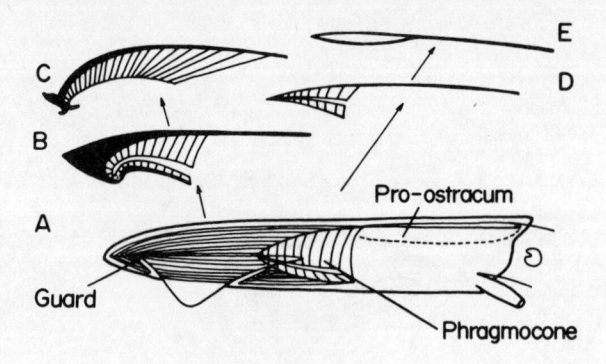

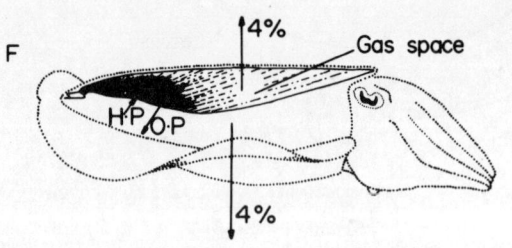

Fig. 25.7 Evolution and reduction of shell in Coleoidea. A. Belemnoid with complete internal shell. B. Primitive sepioid *Belosepia*. C. *Sepia*. D. *Conoteuthis*. E. *Loligo* (A—E after Morton & Yonge). F. Diagram summarizing knowledge of the cuttlebone. The cuttlebone here represented has a density of about 0.6. Liquid within the cuttlebone is marked black. It can be seen that the oldest and most posterior chambers are almost full of liquid. If they were filled with gas this would tend to tip the tail of the animal upwards. The newest 10 or so complete chambers, which lie centrally along the length of the animal, are completely filled with gas. These chambers can give buoyancy without disturbing the normal posture of the animal. The hydrostatic pressure (H.P.) of the sea is balanced by an osmotic pressure difference (O.P.) between cuttlebone liquid and the blood. In sea water the cuttlebone gives a net lift of 4% of the animal's weight in air and thus balances the excess weight of the rest of the animal (after Denton).

continually-replaced radula bearing many teeth (Fig. 25.8) and a long gut with chitinases and cellulases. Predatory molluscs are usually highly mobile or specially adapted for prey capture in some dramatic way (e.g. the proboscis of neogastropod prosobranchs, the poisoned radula barbs of *Conus*). The radula may be powerful, but the gut is shorter with a retentive stomach which stores the intermittently taken food. Cephalopods (squids) use two long tentacles to capture prey fixated visually at a given distance in front of the animal. Scaphopods are microphagic predators using the captacula around the mouth to pick up particles and small animals for transport to the mouth.

Suspension feeders possess large ctenidial surfaces which sieve water. Mucus is copiously secreted and sweeps up material from the ctenidial surfaces towards the labial palps and thence to the mouth. A crystalline style (containing amylase) is usually present as it is in other herbivores (Fig. 25.9). The rectum is usually long and the anus opens into the mantle cavity.

Osmoregulation/excretion

a Excretory organs are developed from the coelomic surface covering the heart, the pericardium, the walls of which may be glandular (Gastropoda, Cephalopoda, Lamellibranchia).

particulate matter. A few amphineurans, many prosobranch gastropods and all cephalopods are predatory and carnivorous. Lamellibranchs and some gastropods (e.g. *Crepidula*) are suspension feeders.

b Grazing algal feeders (e.g. *Patella*) are well provided with a strong buccal mass, a lengthy and

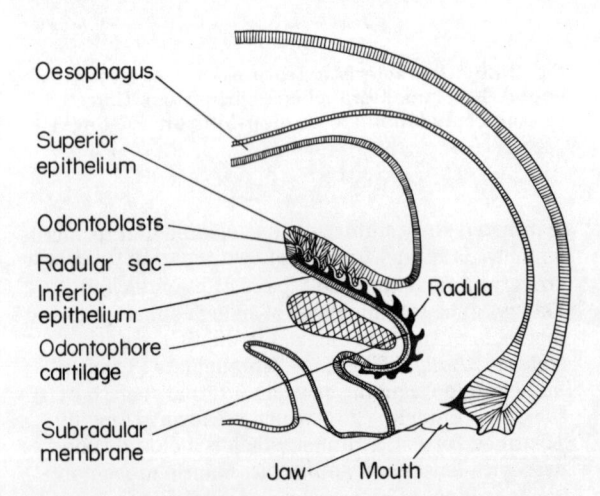

Fig. 25.8 Longitudinal section of a gastropod head showing radula (after Runham).

144   Phylum MOLLUSCA

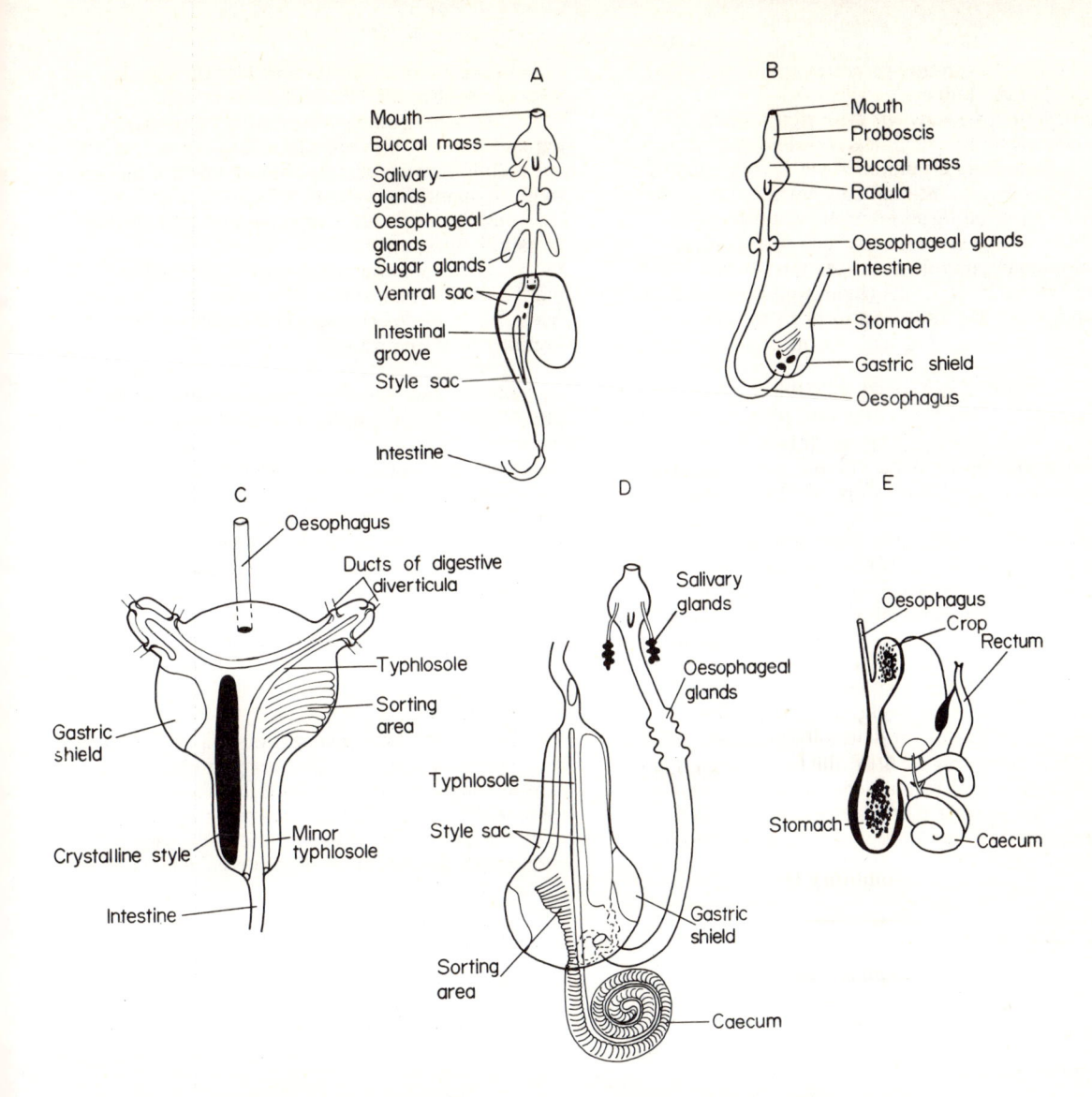

Fig. 25.9 Digestive system of A. a polyplacophoran (Amphineura). B. a scaphopod. C. a lamellibranchiate (eulamellibranch). D. a primitive gastropod. E. a cephalopod (*Octopus*) (A—D after Owen; E after Bidder).

Coelomoducts opening from the pericardium may also be glandular and form a renal organ or kidney (Amphineura, Gastropoda, Cephalopoda), which opens to the exterior via a pore and to the interior via the reno-pericardial pore.

b The blood in the heart of molluscs has a high hydrostatic pressure. A clear filtrate passes through the walls of the auricles and ventricle into the pericardium from whence it passes into the kidney proper. Osmoregulation occurs by resorption of ions in the kidney, excretion by nitrogen moving into the fluid of the kidney.

Marine molluscs may show limited powers of osmoregulation but fresh-water pulmonates show considerable daily clearance of water, maintaining a relatively constant internal concentration of body fluids. Terrestrial forms conserve water to resist desiccation, excretion is uricotelic (uric acid). Ammonotelic (ammonia) or ureotelic (urea) excretion occurs in forms where water is abundant in the habitat. Uricotely is also demonstrated by a few prosobranchs living high on the shore.

Movement

a The normal movements exhibited by representative species are shown in Table 25.1.

b Larvae often move using cilia, but upon settlement other organs come into play. The foot is muscular with peristaltic waves of contraction passing along it in chitons and gastropods. Creeping progression occurs in these groups, and in some lamellibranchs, such as dislodged *Mytilus* (which is normally static). Slow walking on solid substrates also occurs in some cephalopods in which the tentacles provide motive power. The suckers provide adhesion. Burrowing by foot movement occurs in lamellibranchs and scaphopods, the foot changing shape from elongate (for penetration) to anchor-shaped (for purchase to draw the body down into mud or sand).

Swimming takes place in a variety of ways in gastropods (Fig. 25.10), cephalopods and lamellibranchs. Some gastropods (pteropods) expand the foot into wing-like projections that flap the animal through water; others (opisthobranchs) swim by mantle movements. Bivalves swim by clapping the shell valves together and cephalopods by gentle pumping of water from the mantle (slow movement), by rapid ejection of water via the mantle (swift rearward escape) or by fin movements (as in *Sepia*, *Sepiola*). Some squid even fly by jumping from the water.

c Molluscs have some peculiarities of muscular physiology. For example, slow movements are involved in walking and also in shell closure (lamellibranchs); in the latter the physiology of the byssus retractor muscles and other slow adductors has been much investigated. They are capable of prolonged contraction with low energy consumption. Fast movement is the function of muscle with different properties and is found in swimming forms.

Co-ordination

a *i* Nervous system organization (Fig. 25.11) ranges from the very simple to the very complex. Amphineura and primitive molluscs show few ganglionic concentrations, possessing only long cords extending the length of the body with circumoesophageal cords anteriorly. Pedal cords run

Table 25.1 Range of locomotory types exhibited by molluscs, √ indicates occurrence; — indicates no report.

	Creeping	Burrowing	Swimming	Respiratory rhythms	Ciliary
Monoplacophora	√	—	—	—	—
Amphineura	√	—	—	—	larva
Gastropoda	√	sometimes	pteropods, heteropods, some opisthobranchs	pulmonates	larva
Lamellibranchia	—	√	some species	√	larva
Cephalopoda	octopods	—	decapods	√	—
Scaphopoda	—	√	—	—	larva

146 Phylum MOLLUSCA

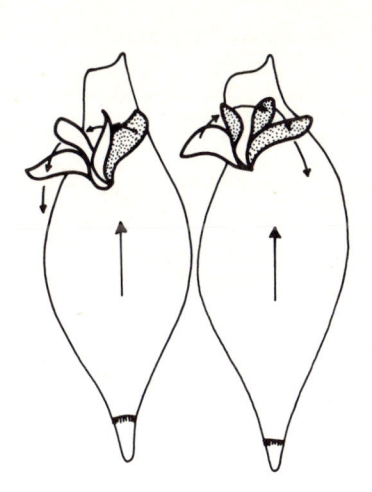

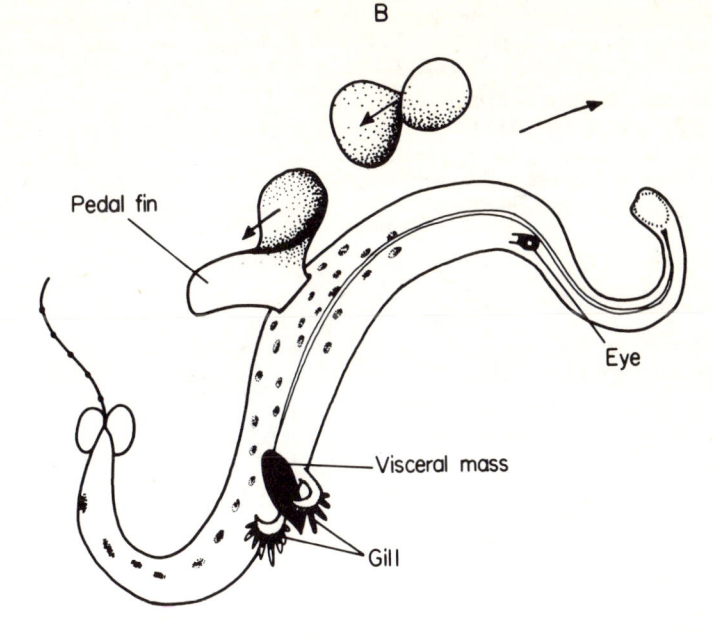

Fig. 25.10 Swimming modes as exhibited by two gastropods. A. A pteropod (*Clione limacina*), seen in lateral view. The parapodia (foot) act as propellers moving ventrally and dorsally alternately (see direction of arrows). Movement is upwards as drawn. B. A heteropod (*Pterotrachea coronata*) in which propulsion is brought about by a median fin (foot) which undulates rapidly. The viscera hang beneath the body and the animal has no shell (A and B after Morton).

to the foot, pallial cords to the mantle. Paired ganglia are found in all gastropods and the full complement is: pleurals at the origin of the pallial cords, the parietals on the visceral loop, the visceral(s) where the nerve loop crosses the gut, the cerebrals anteriorly, the pedals serving the foot, and buccals associated with the buccal mass. There may be other collections of nerve cells associated with particular organs. Lamellibranchs have simple systems as do scaphopods.

The cephalopod nervous system is greatly concentrated and homologies with other molluscs are not readily found. Fig. 25.12 shows details. The stellate ganglia (not shown) act as relay stations of the final motor pathways innervating the mantle musculature. Much work has been carried out on the behaviour of cephalopods.

Sense organs are found as shown in Table 25.2.

ii Work on the physiology of the molluscan nervous system has concentrated on the value that large cell-size has for the investigator. *Helix*, *Pleurobranchea*, *Aplysia* giant cells, and *Loligo* giant fibres have been much used. Details of function are too numerous to give, but pathways, axon distributions, conduction velocities, synaptic phenomena, micro-pharmacology and transmitters have all been studied.

b
Neurosecretion has been described in molluscs and other tissues have also been shown to possess endocrine attributes. Endocrine organs are not well known but the optic gland of *Octopus* governs sexual maturation. Juxta-ganglionic tissue (of nervous origin), e.g. the vena cava organ, may be hormone-producing.

Metabolic function, regeneration and growth, hibernation (e.g. snails), egg-laying and other phenomena may be controlled by endocrine factors. hibernation (e.g. snails), egg-laying, sexual maturation, yolk production, and heart function are all probably controlled by endocrine factors.

Phylum MOLLUSCA 147

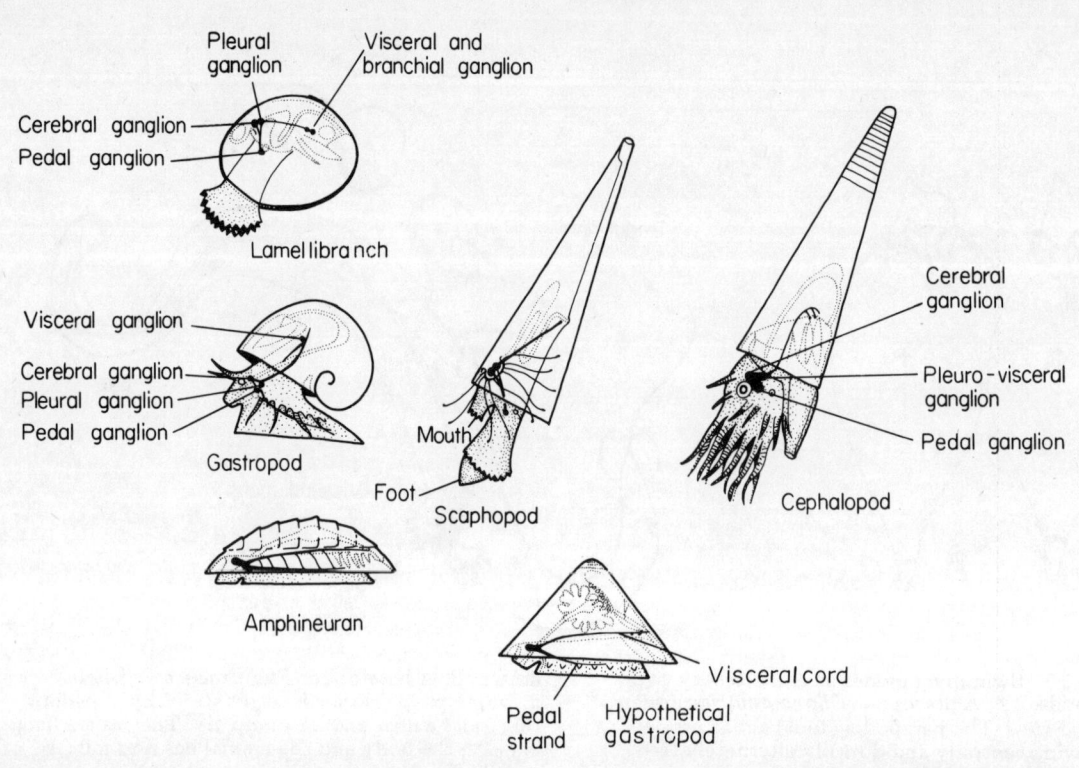

Fig. 25.11 Scheme of molluscan types, showing the plan of the nervous system. One ring around the gut is completed by the cerebral commissure above and pedal commissure below. The second is formed by the cerebro-pleural connective and the pleuro-visceral connective (or visceral loop) and completed by the cerebral and the visceral commissures (or in place of the latter by the fusion into a single visceral ganglion). There is typically a pleuro-pedal connective, but usually no direct connection between pedal and visceral ganglia (except in Amphineura) or cerebral and visceral ganglia. The gastropod shown is a prosobranch with torsion. The cephalopod is a tetrabranchiate with several kinds of tentacles (after Naef).

Table 25.2 Sense organs of molluscs. √ indicates presence; — indicates absence.

	Eyes	Statocysts	Osphradium	Tactile	Chemo-receptors
Monoplacophora	?	—	—	√	probable
Amphineura	megal-aesthetes	√	—	√	√
Gastropoda	tentacles	√	√	√	√
Lamellibranchia	mantle	√	√	√	√
Scaphopoda	—	√	—	√	√
Cephalopoda	complex, retinal	√	—	√	√

148   Phylum MOLLUSCA

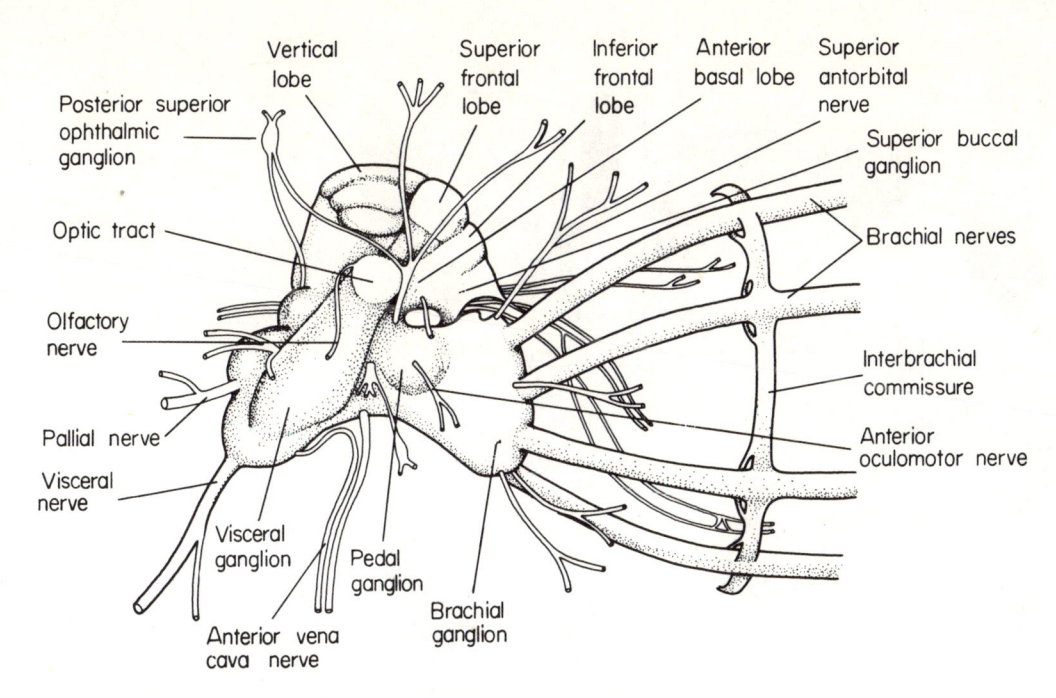

Fig. 25.12 The octopus brain from the lateral aspect to show ganglia (after Young).

Respiration

a The respiratory surfaces are the gills or ctenidia, with few exceptions. In most molluscs these are protected by the mantle and the shell. In Amphineura there may be many ctenidia along the sides of the body, lying in a channel, but in Aplacophora the whole system is greatly reduced to a small pumping mantle cavity with little in the way of gill structures.

Neopilina bears five pairs of gills. Prosobranchs may have two ctenidia (e.g. *Haliotis*) or more commonly one (following torsion); opisthobranchs develop new gill-like structures and the ctenidia are lost; pulmonates are air-breathing with a vascularized lung (mantle), except in fresh-water forms where secondary gills are developed. Lamellibranchs have ctenidia of greater or lesser complexity, subserving a feeding purpose. In cephalopods the ctenidia are four (in tetrabranchs) or two (dibranchs) and lie in the mantle cavity.

b Water from the habitat is circulated across ctenidial surfaces by a variety of means. Ciliary beating moves water directly over the surface of the ctenidium in all groups, but larger currents are generated by muscular means. Lamellibranchs pump using adductor muscles; ventilation is by opening and closing the pneumostome in pulmonates; and the mantle muscle provides the impetus for moving water in and out of the mantle in cephalopods, a one-way stream being achieved by the arrangement of valves at the funnel opening.

Circulation/coelom

a In most forms the coelom is represented only by the pericardium, kidneys and the gonadal cavity (Fig. 25.13), connecting to the exterior via coelomoducts. These are most numerous in *Neopilina* where there are 6 kidney ducts and 2 pairs of gonads. In cephalopods there is a large and

Phylum MOLLUSCA

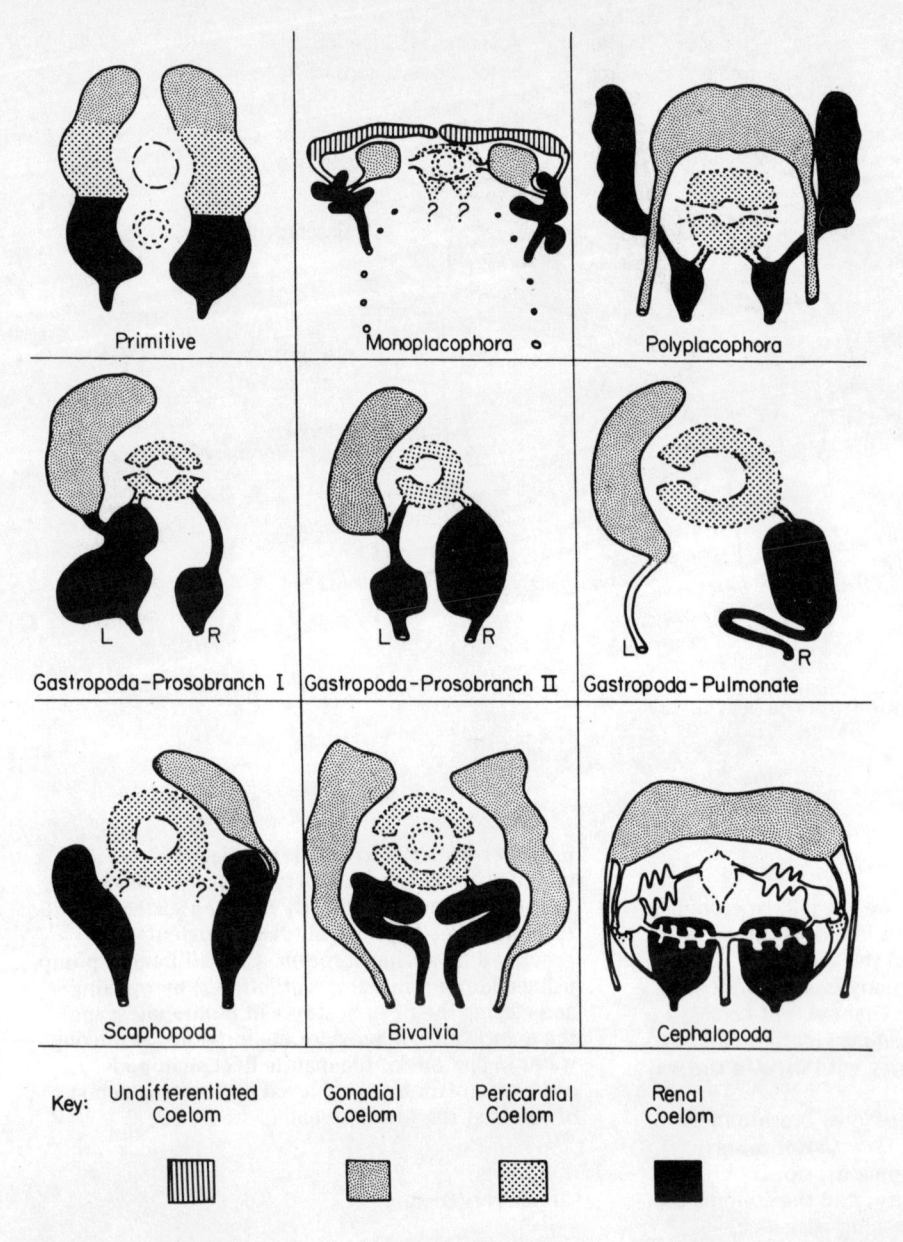

Fig. 25.13 Relationships of the gonadal, pericardial, and renal coelomic derivatives in the phylum Mollusca. The representation for the Monoplacophora is that for *Neopilina*. The line Prosobranch I leads to conditions in extant archeogastropods except the Neritacea. The line Prosobranch II leads to conditions in the Neritacea and in the meso- and neogastropods. The representation for the Cephalopoda is that of *Octopus* (after Martin & Harrison).

150   Phylum MOLLUSCA

true coelom in which the heart lies (except octopods) but it bears a similar relationship to pericardium and kidneys.

Other body cavities are filled with blood and form sinuses, these being expanded portions of the blood system. Blood circulates from the heart via an aorta from the ventricle, but then leaves the closed system, permeates the body tissues, and eventually returns to the sinuses and thence to the ctenidium, returning to the heart via the kidney. Cephalopods are the only group to possess arteries, and the haemocoelic system is absent. The haemocoel in the other classes forms a hydrostatic skeleton, especially valuable for burrowing and locomotory activities.

b Propulsion of the blood is achieved by two major means, (i) a large muscular heart (absent in Scaphopoda and a few unusual gastropods), and (ii) movements of the body wall and the other organs of the body. The heart has a powerful ventricle and may have one or two auricles depending upon class and species. Blood generally flows directly to the anterior nervous system and returns via the system of spaces discussed above. In species where the distension of the body wall is achieved by hydrostatic means the contraction of muscles within the organs is sufficient to move blood around the internal haemocoel. Valves prevent backflow from sinus spaces.

c The respiratory pigment is haemocyanin, containing copper. The properties of gastropod blood allow a greater use of oxygen circulating in the mantle cavity (56% in *Haliotis*) than in lamellibranchs (5—9%) but less than in cephalopods (63% in *Octopus*). These figures may be correlated with the small gill and low volume cleared by gastropods, compared with high currents but sedentary habit and low metabolic demand of bivalves, and high respiratory volume and high metabolic requirement of cephalopods.

Gastropod blood is saturated at low oxygen pressures whilst the very active cephalopods reach blood saturation only at higher oxygen pressures. Dissolved oxygen accounts for about 3% of total blood volume in all except cephalopods where it is much higher. Lamellibranchs may have no haemocyanin, oxygen being transported in simple solution, but a few species of bivalves and gastropods have developed haemoglobin as a carrier, this feature being perhaps correlated with the poorly-oxygenated nature of the habitat. There are a number of kinds of amoebocytes within the blood system. Some molluscs have myoglobin in their muscles.

Reproduction

a *Neopilina* is dioecious, possessing two pairs of gonads, that discharge via coelomoducts that double as excretory canals. Chitons are of separate sexes with a single median gonad shedding gametes via separate gonoducts but some of the Aplacophora are hermaphrodite, shed gametes through renal ducts and secrete egg shells. Gastropods have a single gonad, releasing gametes via the renal organs in primitive forms, but having specialized gonoducts in higher groups. These can be glandular and yolk and shell material may be laid down around the egg. *Crepidula* starts life as a functional male, later becomes hermaphrodite and then lastly a functional female. Hermaphroditism is also found in pulmonates and opisthobranchs. Lamellibranchs have simple reproductive systems with paired gonads, short ducts and no glands. The sexes are separate, as they are in cephalopods. In the latter group there is a single median gonad that opens into the coelom and the gametes exit via a coelomoduct that opens into the mantle cavity. There are various glands associated with the production of the yolk (the eggs are large), the shell and the pigment.

b Molluscs exhibit a wide variety of types of reproduction without the introduction of asexual methods. Many shed gametes into the plankton (especially lamellibranchs and some gastropods) where fertilization occurs, economy being achieved by synchrony of the act of shedding throughout a population (a phenomenon they have in common with other phyla). This may be due to maturation under hormonal influence. Some bivalves brood the young, especially fresh-water forms where a larval stage, the glochidium, is released to pass a period as a parasite on fish, attaching itself by small valve teeth or byssus threads. Chitons lay many eggs, externally fertilized and giving rise to planktonic short-lived larvae. Gastropods may lay single strings or capsules of eggs and in neogastropods fertilization is internal. Pulmonates also show elaborate copulation and internal fertilization. Some gastropods show self-fertilization and brooding within a uterus occurs in some. In many aquatic gastropods there is a planktonic larval stage, a trochophore, giving rise to the characteristic veliger. Cephalopods may lay very few eggs or clusters at a time. Reproductive behaviour is complex. Sperm are collected into spermatophores and transferred to the female via a special arm, the hectocotylus, which may be autotomised (*Argonauta*). Development within the egg capsule produces a juvenile form that is planktonic even among the benthic octopods.

References

Morton J.E. 1958. *Molluscs*. Hutchinson University Library, London.

Wilbur K.M. & Yonge C.M. (Eds.) 1966. *Physiology of the Mollusca*, 2 vols. Academic Press, New York and London.

26 Phylum PHORONIDA

Only 15 species known, up to 10 cm in length.

Characteristics

1 Exclusively marine in habitat.
2 The larval type is the planktonic actinotrocha.
3 The adults are tubicolous, either producing a secreted chitinous tube, or boring into mollusc shells.
4 They are elongate, cylindrical, coelomate animals exhibiting bilateral symmetry.
5 Anteriorly there is a terminal lophophore in the shape of a horseshoe.
6 The gut is U-shaped, mouth and anus being close together.
7 There is a pair of metanephridia doubling as gonoducts.
8 A closed circulatory system is present.

Larval form

The larva is planktonic and is known as the actinotrocha (Fig. 26.1). A pre-oral hood overlies the mouth, whilst tentacles fringe a post-oral ridge. The primary locomotory organ is the telotroch, a ring of cilia on a ridge surrounding the anus. Tentacles are added successively to the post-oral group. The actinotrocha may swim for several weeks.

Metamorphosis

Drastic podaxonic rearrangement of the body occurs at metamorphosis. This is said to be a rapid process taking place when the larva becomes benthic. The metasome pouch is protruded through the body wall by muscular contractions and develops axially. It carries the gut into a characteristic U-shape (Fig. 26.2). Much of the larva disappears, and few larval structures are

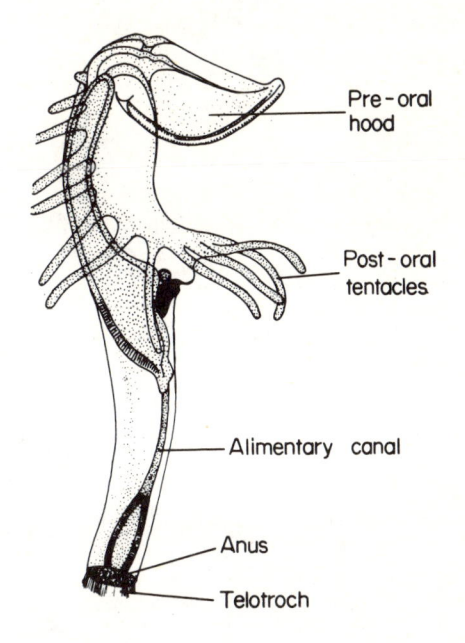

Fig. 26.1 The actinotrocha larva (after Selys-Longchamps).

incorporated into the adult. Tube building begins following metamorphosis and the adoption of a sedentary life.

Adult body form e.g. *Phoronis, Phoronopsis.*

Phoronids are vermiform with an anterior crown of tentacles, a long slim body and a somewhat expanded posterior end (Fig. 26.3). The hollow, ciliated tentacles form a single row on each of the two sides of a double ridge that is thrown into a crescent or horseshoe shape (lophophore) (Fig. 26.4). The mouth is situated between the two ridges and is covered by a flap, the epistome. The anterior

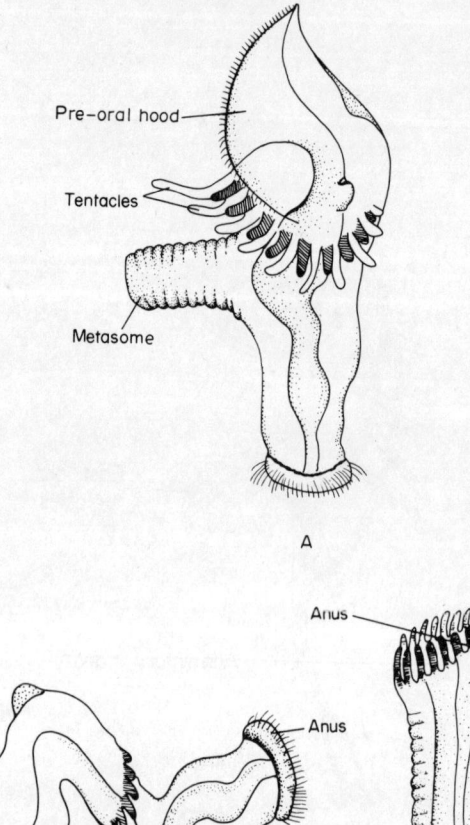

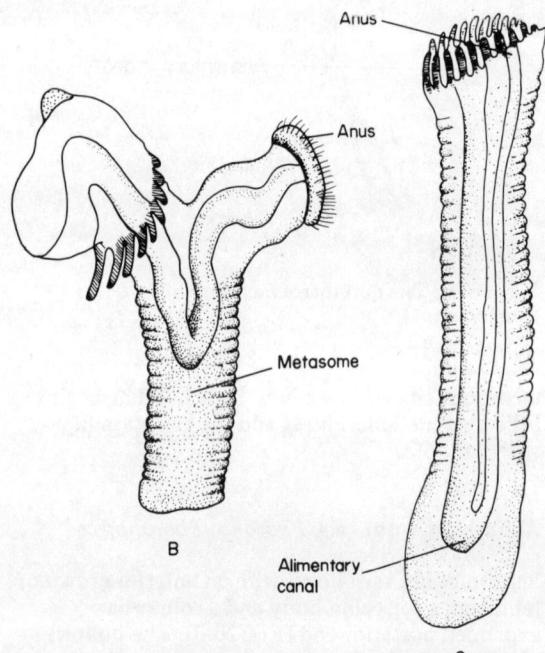

Fig. 26.2 Stages in metamorphosis of the larva. A. Commencement of metasome outgrowth. B. Beginning of gut looping. C. Juvenile type (A—C after Meeck).

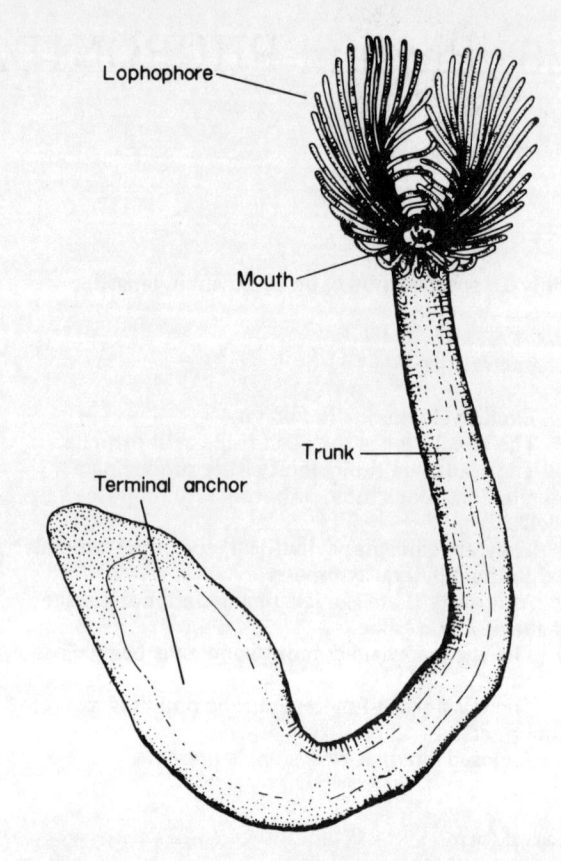

Fig. 26.3 Young adult of *Phoronis architecta* with lophophore and terminal broad anchor (after Wilson).

portion of the animal forms a brood pouch and is bounded dorsally by the anus and two nephridiopores, which are separated from the mouth by the lophophoral ridge. The coelom is represented in two portions. The secreted chitinous tube acts as an external skeleton and the coelomic fluid as a hydrostatic internal one. Tentacles are supported by a thick, rigid basement membrane.

Feeding

a   Phoronids are microphagic, ciliary filter-feeders.

b   The gut is U-shaped (Fig. 26.5). The lophophore is heavily ciliated and water is drawn between the

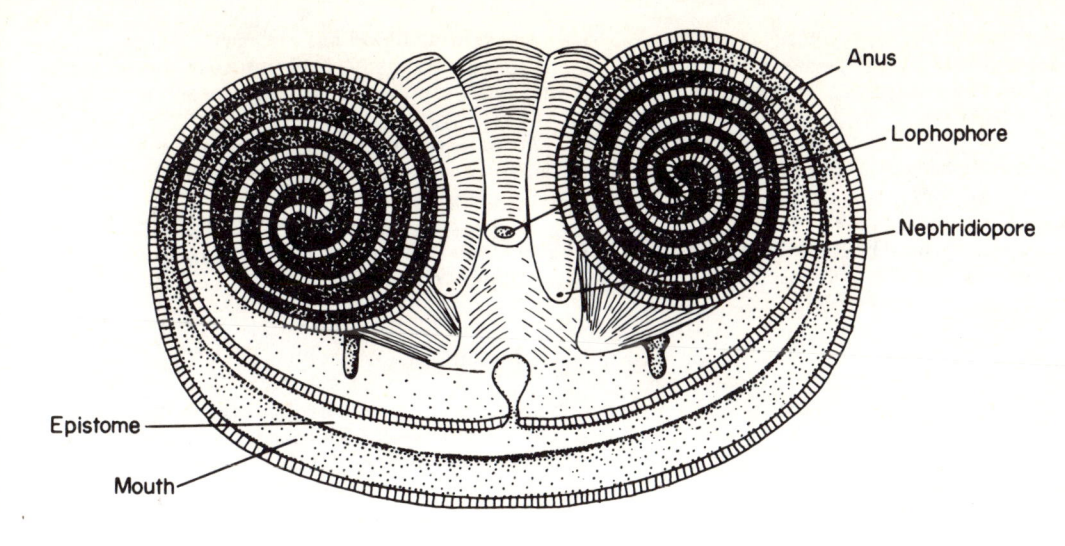

Fig. 26.4 Section of the lophophore of *P.australis* to show spiral arrangement of tentacles (after Benham).

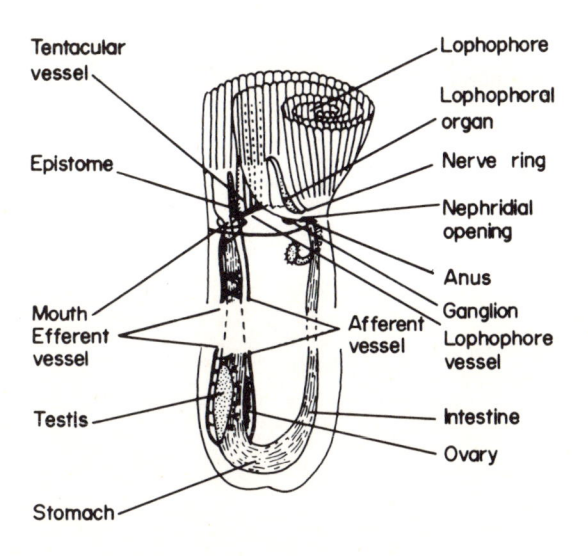

Fig. 26.5 Arrangement of major organs of *Phoronis* (after Delage & Herouard).

tentacles. Grooves along the inner faces direct the current towards the mouth and subsequently over the anus. Mucus is secreted to engulf small particles.

Osmoregulation/excretion

a   The larva possesses protonephridia, but the adult develops metanephridia. There is one pair, each being U-shaped, totally ciliated and opening into the metacoel, sometimes by two funnels. The external openings or nephridiopores are near the anus.

b   The metanephridia are presumed to be excretory. The nephridiopores exude fluid whose composition is not known but uric acid has been reported present. The physiology of osmoregulation is not known; it seems likely that phoronids are osmotic conformers.

Movement

a   Movements are stereotyped, being restricted to locomotion within a tube. Extension of the lophophore takes place in undisturbed conditions, rapid withdrawal follows vibration or a touch,

Phylum PHORONIDA   155

especially on the base of the lophophore. Tentacles may move independently. Some species burrow within eroded shells, and others can burrow if removed from the tube.

b Muscle is arranged in a typical vermiform manner with a thin, outer layer of circular fibres and a thicker, inner, longitudinal array. The fibres are of smooth muscle, and may be syncytial. Sphincters are present in some species. Tentacles are not heavily muscularized, nor is the lophophore.

c Physiology is not known.

Co-ordination

a *i* The body wall is reported to be heavily innervated, with many sense cells, association and motor fibres situated close to the base of the epidermis. Anteriorly this nerve layer is thickened as a ring along the outer ridge of the lophophore. From this arise tentacle nerves. Only one long 'nerve' exists, a giant fibre, restricted usually to the left side only. Only one sense organ, associated with the lophophore, has been described.

ii The epidermal nerve net conducts equally and decrementally in all directions. Rapid contractions are mediated by the giant fibre.

b No hormones have been described.

Respiration

a No special respiratory organs have been described.

b Respiratory exchange presumably takes place across all parts of the body surface, especially the tentacles.

Circulation/coelom

a The coelom is bipartite, there being a trunk metacoel with metanephridial coelomoducts, and a lophophore mesocoel with no connections to the exterior. These are separated by a septum. The well-developed blood system is closed. Two main longitudinal trunk vessels (dorsal, blood going forward; ventral, blood passing rearward) communicate via many small sinuses in the gut wall. A ring vessel supplies the lophophore and tentacles.

b All vessels seem capable of contraction at independent rates. The long vessels provide the major propulsive power.

c The blood is colourless but red corpuscles (with haemoglobin) circulate in it.

Reproduction

a Phoronids are mostly hermaphrodite but some species are dioecious. The gonads are located in the coelom on the peritoneal wall as ill-defined aggregations of cells. In hermaphrodites the testes and ovary occupy different positions.

b *Sexual reproduction*
Gametes are shed into the coelom and escape via the nephridia. Fertilization may occur in the coelom or in sea water. Some species brood their eggs and embryos.

c *Asexual reproduction* by transverse fission has been described for *Phoronis ovalis*. Autotomy and regeneration of parts has also been reported.

References

Grassé P.P. (Ed.) 1959. *Traité de Zoologie*, Vol. V (1). Masson et Cie, Paris.
Hyman L.H. 1959. *The Invertebrates*, Vol. V. McGraw-Hill, New York.

27 Phylum ECTOPROCTA (=BRYOZOA)

Approximately 4,000 species extant. Colonies may reach several cm, individual zooids microscopic.

Characteristics

1   Aquatic, mostly marine but some estuarine or fresh-water species.
2   Sedentary, colonial, composed of zooids.
3   Possess a lophophore of post-oral tentacles which are retractable into a sheath.
4   Coelomate.
5   Zooids secrete rigid or gelatinous walls.
6   Gut is U-shaped, anus opens close to mouth but external to lophophore.
7   Colonies are hermaphrodite.

Larval form

Electra, *Membranipora* and some other ectoprocts (bryozoans) that shed eggs directly into the sea have a characteristic triangular larva, the cyphonautes. This is bivalved and possesses a tuft of cilia that projects from the apex of the triangle, whilst the base is fringed by a circular border of motile cilia (Fig. 27.1). It possesses an alimentary canal.

Other marine bryozoans have a larval stage that is unnamed (Fig. 27.2A); it is frequently brooded in special sacs within the zooid or colony. There is great ciliation of the surface of the larva. There is no gut.

Fresh-water bryozoans have a complex development that leads to a 'larva' that is effectively a juvenile colony.

Metamorphosis

The larva usually settles rapidly after release, having selected an appropriate site. An adhesive sac secretes material that acts as a cement for attachment. The

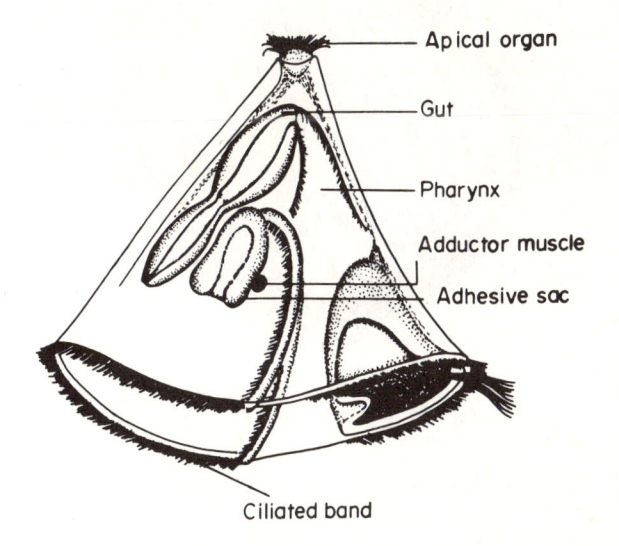

Fig. 27.1   Cyphonautes larva (after Ryland).

apical part of the larva then moves downward in an 'umbrella' stage (Fig. 27.2B). All larval organs are then resorbed. The metamorphosis is complete with the formation of the first zooid or ancestrula.

Adult body form

The Ectoprocta are colonial and many individual zooids exist close together in one mass. Whilst there are basic similarities throughout the colony there is also polymorphism, with specialization of a variety of individuals, in some examples.

Zooid types

i   Autozooids: normal individuals carry a lophophore (Fig. 27.3) which can be retracted or everted.

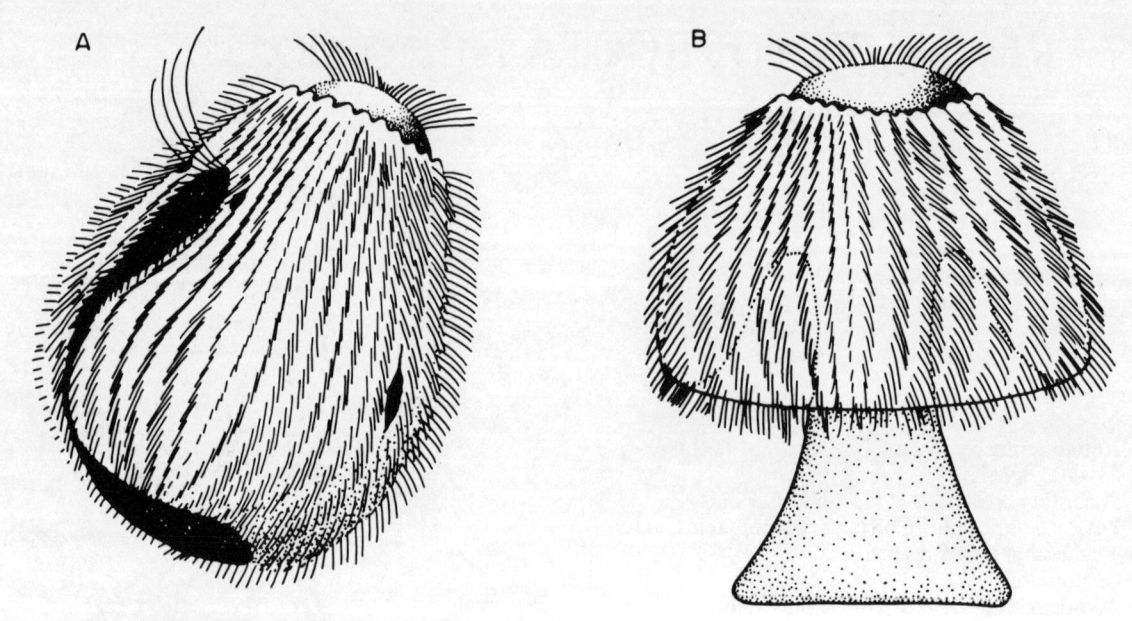

Fig. 27.2 A. Ciliated larval type (*Bugula*). B. Umbrella stage of metamorphosis of *Bugula* larva (A and B after Ryland).

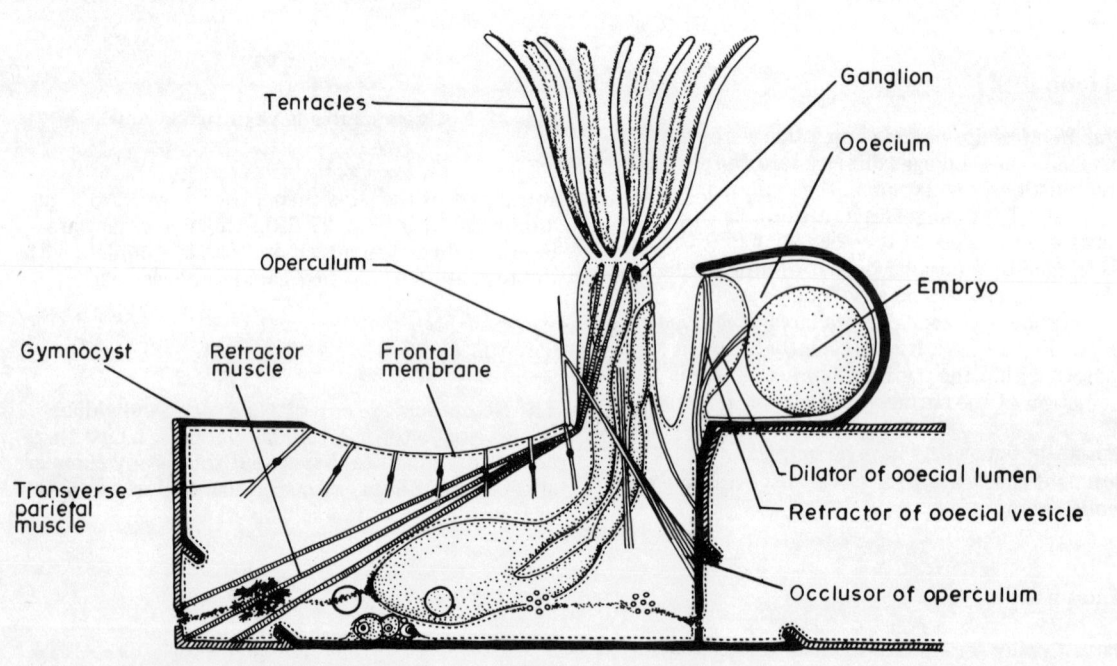

Fig. 27.3 Normal individual bearing an ooecium or brood chamber on the anterior border (after Ryland).

158   Phylum ECTOPROCTA

ii Avicularia (Fig. 27.4A): the operculum is greatly altered, hinged at the base but elongated and tapered to form a mandible that closes into a space termed the palate. There are paired muscles to effect closure upon appropriate mechanical stimulation.

iii Vibracula (Fig. 27.4B): individuals with an elongate bristle or seta that is articulated in a pair of asymmetrical condyles which allow movement around the axis of the seta. They are moved by abductor and adductor muscles.

iv Kenozooids: zooids almost devoid of internal structure, found as components of stolons, stalks, rhizoids and other attaching organs.

v Ooecia (Fig. 27.3): modification of an otherwise normal individual to form a brood chamber either on the outside or between two cuticular surfaces.

There are three classes.

1 *Class* PHYLACTOLAEMATA e.g. *Cristatella*, *Plumatella*.
Zooids cylindrical with horseshoe-shaped

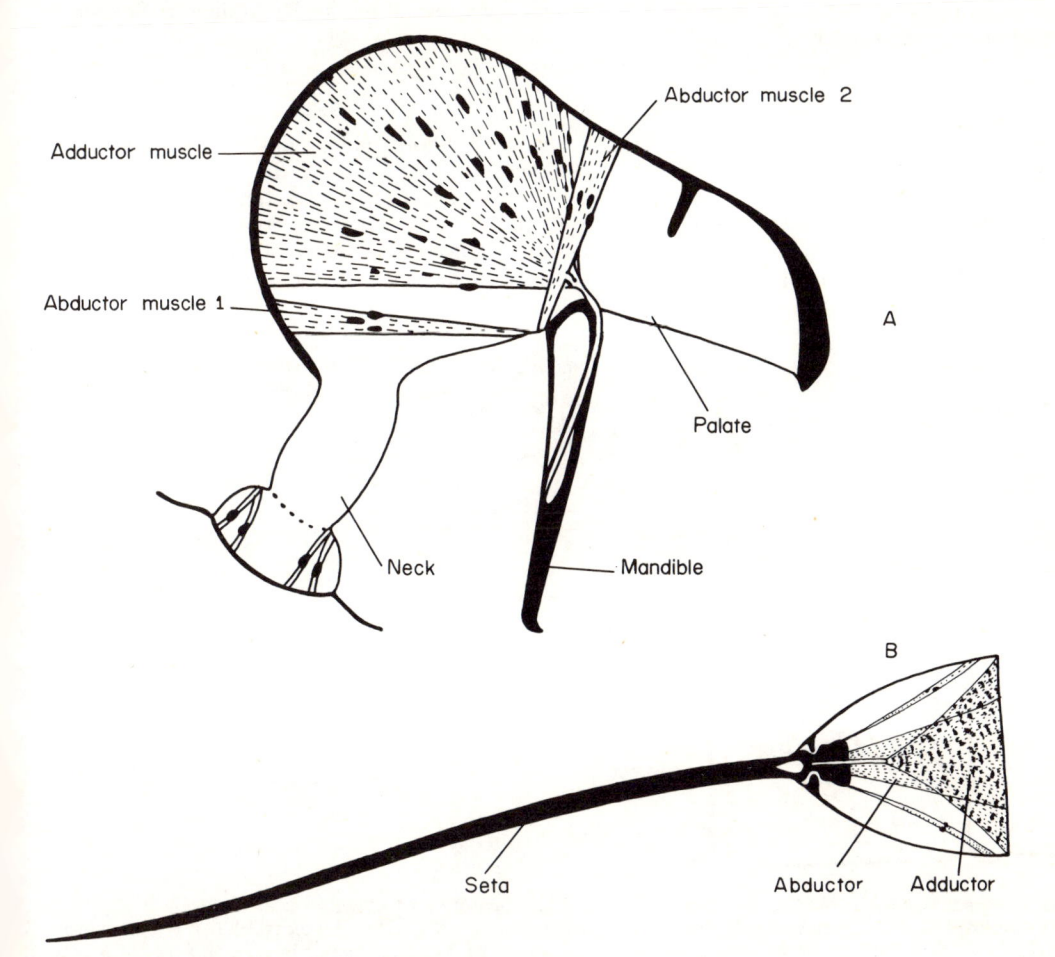

Fig. 27.4 Representative polymorphic individuals. A. Avicularium. B. Vibraculum (A and B after Ryland).

Phylum ECTOPROCTA

lophophore; colonial walls non-calcified; coelom continuous between individuals; produce statoblasts; fresh-water.

2 *Class* CTENOLAEMATA e.g. *Crisia, Cyclopara*.
Zooids cylindrical; some calcification; coeloms of zooids separated by septa; limited polymorphism; marine.

3 *Class* GYMNOLAEMATA e.g. *Membranipora, Alcyonidium*.
Zooids cylindrical or squat with circular lophophore; some calcification; coeloms separated by septa, communicating by pores plugged with tissue; exhibit polymorphism; marine.

Feeding

a Ectoprocts are suspension feeders.

b They filter planktonic entities from the watery environment by means of the tentaculate lophophore. Water is drawn in at the top of the bell and passes out between the tentacles. Material is ingested by a suction pump method. The gut is U-shaped, the mouth opening within the ring of tentacles, and the anus outside it. Food is digested and absorbed in the stomach and is continually rotated by cilia as a cord.

Osmoregulation/excretion

a There are no special adaptations.

b Some excretory products may be removed via brown body formation (see Reproduction Section)

Movement

a The lophophore indulges in slight side-to-side exploratory movements, but the most obvious activity is the very rapid retraction of the lophophore when touched or when vibration occurs. Protraction occurs by hydrostatic means (Fig. 27.5). Avicularia snap at touching objects.

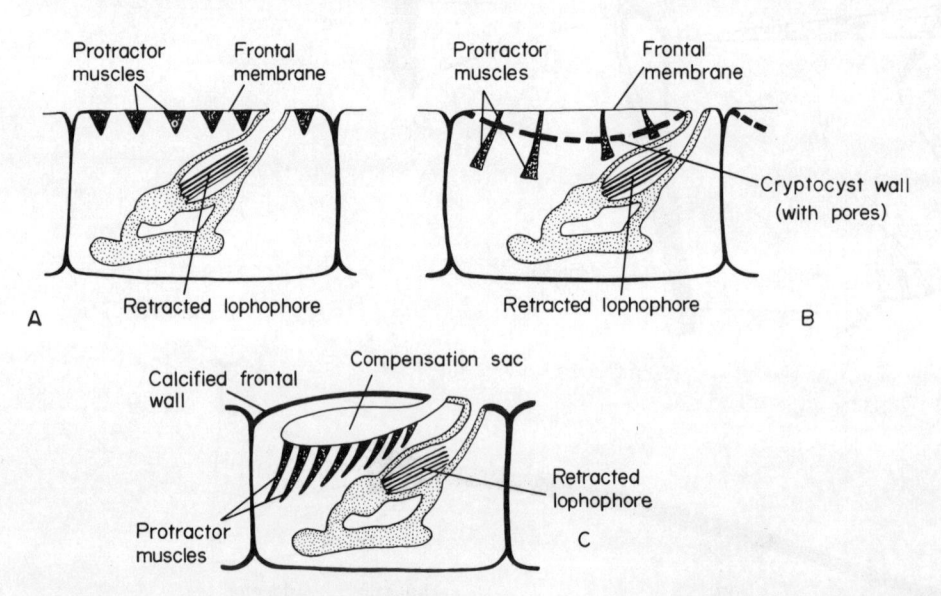

Fig. 27.5 Mechanisms involved in protraction and retraction of the lophophore. A. ANASCAN grade: the frontal membrane is flexible and when drawn in by protractor muscles internal hydrostatic pressure rises and the lophophore extends. B. Also ANASCAN: the cryptocyst wall provides protection but the principle of eversion is as in A. C. ASCOPHORAN grade: the exoskeleton is complete and water movements take place in and out of a compensation sac or ascus under the influence of the musculature (A, B and C after Russell-Hunter).

b Fast muscle, innervated by rapid conduction pathways, seems to be indicated.

Co-ordination

a There is a small CNS with, sometimes, a dermal nerve network. The ganglion is dorsal and one pair of sensory nerves and one pair of motor fibres pass into each tentacle. It is not clear if there is a colonial nerve system connecting all zooids to one another.

Sense organs present include mechanoreceptors (touch sensors) on the tentacles and possibly on the avicularia. Larvae are light-sensitive.

b Some developmental evidence exists for a hormone controlling formation of ovicells, but it may also be for co-ordination of growth cycles in the colonies.

Respiration

No special structures exist.

Circulation/coelom

a There is no blood system.
The basic structure of the body is tripartite and this bears some resemblance to the situation in brachiopods and phoronids — two other lophophore-bearing groups.

The anterior coelom or protocoel is represented in the epistome (a flap of tissue overlying the mouth) in Phylactolaemata. The lophophore and its tentacles develop from the mesosome (enclosing the mesocoel) and the major body cavity is the metacoel (in the metasome).

b Physiology is not known.

Reproduction

a Ectoprocts are generally hermaphrodite. The zooids have an ovary and one or more testes developing on the peritoneum lining the coelom. The gonads are simple and ductless.

b *Sexual reproduction*
The gonads shed their products via the coelom and coelomoducts to the exterior. Many species brood the cross-fertilized egg in a specialized compartment, the ooecium.

c *Asexual reproduction*
Growth of the colony and addition of new individuals is a function of a system of stolons or by formation of partition walls across existing zooids to form more chambers (budding).

Periodic renewal of individuals takes place. The tentacles, gut and the whole of the individual degenerate to form a brown body. This is later invested by a newly-grown polypide arising from the main colony, and comes to lie in the gut, eventually being voided through the anus. This phenomenon appears to be a response to poor conditions or onset of winter, or to result from brooding of embryos, or from senescence.

Reference

Ryland J.S. 1970. *Bryozoans*. Hutchinson University Library, London.

28 Phylum BRACHIOPODA

About 300 species extant. Most are rather small, up to several cm in length. There are over 12,000 fossil species.

Characteristics

1 All are marine, the majority inhabiting shallow water.
2 A few burrow, but most are sedentary and attached.
3 There is a free-swimming larva.
4 The body is enclosed in two shell valves arranged dorsally and ventrally, the latter generally being the larger. Attachment may be by a pedicle.
5 The feeding organ is a complex lophophore, which has an internal skeleton in some species.
6 Coelomate.
7 There are no organs of locomotion, no complex sense organs.
8 Most bodily organ systems are simple.

Larval form

Larval development is not well known. Larvae of the articulate group consist of those segments, partly ciliated, that give rise to the body, mantle and pedicle respectively. Inarticulate larvae possess only two segments, the body and mantle progenitors. This type resembles an adult brachiopod.

Metamorphosis

Inarticulates such as *Crania* undergo no metamorphosis, settling on the ventral valve and becoming cemented to the substrate. *Lingula* attaches by the pedicle and grows directly from juvenile to adult stages. Some species have prolonged larval lives. Articulate larvae show a reversal in position of the mantle lobe which first grows around the pedicle, and later around the body segment. Secretion of shell valves begins after this rearrangement.

Adult body form

Brachiopods were once a dominant phylum in the marine fauna. They are now few in number, but of these *Lingula* and *Discona* are extant in virtually the same form as shown in the Cambrian era.

Brachiopods have a superficial resemblance to the lamellibranch molluscs. This is due to the presence of two mantle lobes that secrete two shell valves which become calcified. The plane of symmetry runs through the valves and not between them. The valves are generally termed dorsal and ventral although the actual orientation after settling may be the reverse. The valves are separate but may be hinged at the posterior end. Adductor muscles join the two valves together. The body proper is small within the mantle cavity (Fig. 28.1). A large part of the mantle cavity is occupied by a complex lophophore (Fig. 28.2). This may be supported by a skeleton, the brachidium. Basically a pair of ciliated, tentaculate arms, varying in length and degree of coiling, the lophophore serves as a feeding and respiratory organ.

1 *Class* ARTICULATA e.g. *Terebratulina*. Possessing a hinge of interlocking teeth and sockets between the valves; lophophore with internal skeleton; intestine ends blindly.

2 *Class* INARTICULATA e.g. *Lingula*. The valves are held together by muscles and there is no hinge; lophophore without internal skeleton; anus present.

Feeding

a Brachiopods are suspension feeders.

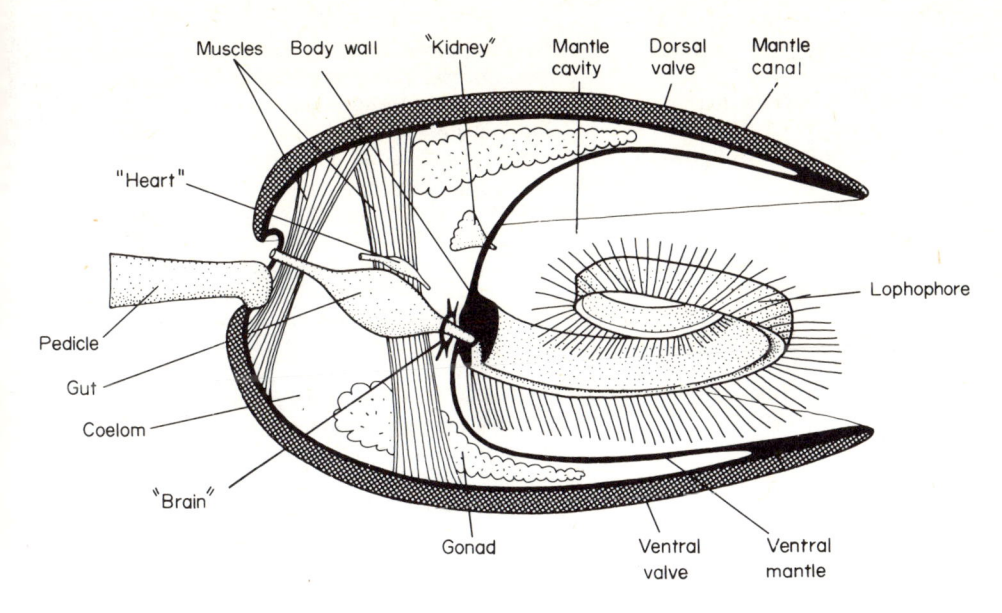

Fig. 28.1 Diagram of the anatomy of a brachiopod (after Rudwick).

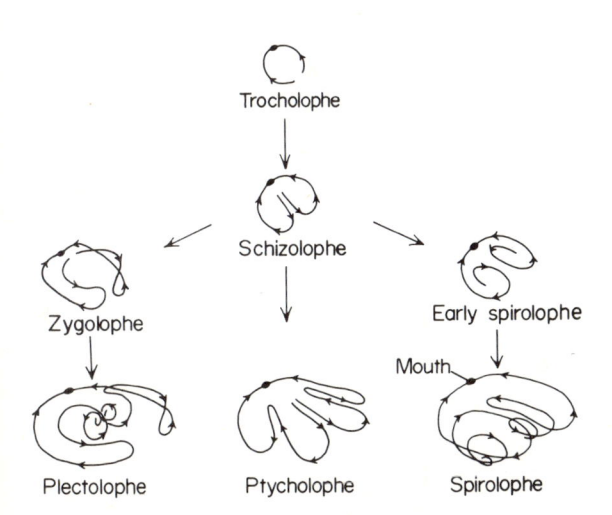

Fig. 28.2 Three alternative lines of lophophore development. Arrows indicate direction of transport of food particles along the branchial axis (after Rudwick).

b The gut is tubular, has ramifying digestive diverticula originating at the stomach, and ends in an intestine. In articulates this is blind ending; inarticulates possess an anus.

During feeding the shell valves gape open and water passes into the enclosed mantle cavity. Circulation is maintained by the activity of the lophophore cilia, and is unidirectional. Water enters an inhalent chamber, moves between the lateral cilia of the lophophore filaments and exits from an exhalent chamber. Mucus removes trapped particles of diatoms and algae and is moved to the mouth. Some selection of particles occurs, and the adductor muscles may close the shell valves rapidly to eject unwanted material and faeces.

Osmoregulation/excretion

a There may be one or two pairs of metanephridia opening internally into the coelom, and externally at a nephridiopore.

b Coelomocytes ingest the excretory material and are then ejected via the metanephridium into the mantle cavity. Nothing is known of osmoregulation in the group.

Phylum BRACHIOPODA

Movement

a When the animal is feeding the shell valves gape, allowing water to move into the mantle cavity under ciliary action of the lophophore. Those brachiopods such as *Lingula* that burrow in a soft, muddy substrate can withdraw from the surface by retraction of the pedicle (Fig. 28.3). In these forms the shell is shallow and parallel-sided, presumably allowing easier withdrawal. Protective reactions in sessile species consist of rapid closure.

b The pedicle is different in structure in the two classes. In articulate species the larval pedicle grows longer, as a cylinder with a core of cartilaginous substance covered by a thick, chitinous cuticle. It is attached to the valves within the coelom by adjustor muscles. Contraction of these brings about various changes in the position of the animal. Inarticulate brachiopods possess a pedicle that develops from the ventral mantle edge. The pedicle is hollow, muscular, and contains a coelomic cavity. There are muscles in the wall or cavity and a sphincter at the connection to the coelom, the whole structure acting as a hydrostatic skeleton. Most brachiopods are attached firmly to a hard substrate, but the chemical nature of the cement is unknown. Few brachiopods root into a soft substrate.

c Very little is known of the physiology and structure of brachiopod muscle, but adductor muscles are formed of two portions that may be fast and slow (or quick and catch) in function as is the case in lamellibranch molluscs.

Co-ordination

a *i* The nervous system is a simple one. The brain is a collection of small nerve cells on the ventral side of a circumoesophageal ring. Lengthy nerves emanate from this area to supply the mantle edge, and are presumably sensory in function. Motor nerves run to the adductor and other muscles (including the pedicle) and to the lophophore. The mantle edge is very sensitive to tactile stimulation, and there are indications that chemical influences are strong but the end organs are not known. There are no eyes or pigment spots or other special sense organs with the exception of statocysts in *Lingula*.

ii There may be fast and slow muscle and motor control systems in brachiopods, since both rapid closure and withdrawal movements have been described as well as slower alterations of valve position. The adductor muscles are reported to possess two anatomically distinct regions. This may be reflected also in the features of the nervous system.

b There are no reports of hormones in brachiopods.

Respiration

a Whilst the lophophore has undoubtedly been developed and adapted as a large, ciliated feeding

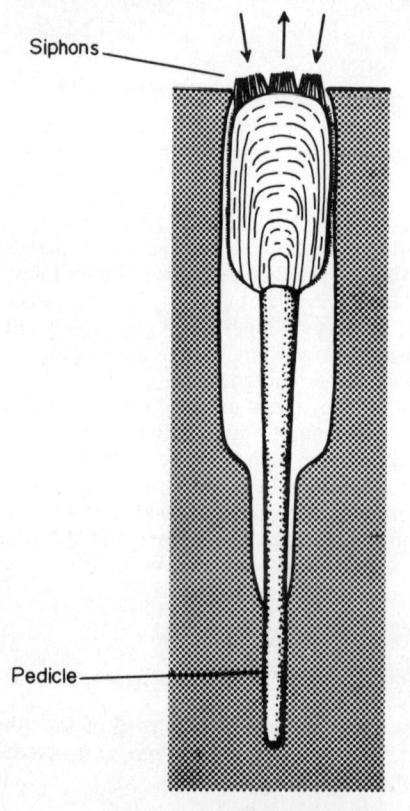

Fig. 28.3 The position of *Lingula* in its burrow. The shell is withdrawn from the surface by contraction of the pedicle. Water enters and leaves the mantle cavity via the three siphon-like openings (after Rudwick).

surface it seems probable that it doubles as a respiratory organ. Respiratory exchange also may take place across the extensive surface of the mantle and body wall. In one example (*Glottidia*) ampullae project into the mantle cavity.

b There is no information on the physiology of respiration in brachiopods.

Circulation/coelom

a The coelom arises in schizocoelic fashion in inarticulates, as cavities in mesodermal masses; in articulates on the other hand, the coelom arises as a pair of sacs (coelomic pouches) growing out of the archenteron, i.e. is of enterocoelic derivation.
   The main coelomic space is large and gives rise to canals that permeate the mantle, ending blindly all around the mantle edge. A series of open-ended, fine vessels ramifies in the mantle canals, and may represent a blood system. There is a heart.

b Coelomic fluid circulates within the canals (see above) of the mantle and provides a circulatory system.
   The physiology of the canals and of the 'blood system' is not known.

c There is no blood pigment reported.

Reproduction

a The sexes are generally separate. There are usually four gonads, arranged two dorsally and two ventrally.

b Ripe gametes are released into the coelom and leave via metanephridia. Fertilization is external. In some species the embryos are brooded in specialized brood pouches. Cleavage is regular and complete, giving rise to a larva.

Reference

Rudwick M.J.S. 1970. *Living and Fossil Brachiopods*. Hutchinson University Library, London.

29 Phylum CHAETOGNATHA

60 species known. Length 1–10 cm.

Characteristics

1 Marine; typical members of plankton except for *Spadella* (benthic).
2 Bilaterally symmetrical.
3 Coelomate, with an enterocoelic development.
4 Body comprises three regions: head, trunk and tail.
5 Possess lateral and caudal fins.
6 Head has a pair of eyes and a set of grasping spines for prey capture.
7 Longitudinal muscles of unusual type, arranged in quadrants. No circular muscles.
8 A vertical septum is present dividing trunk from tail.
9 No circulatory or excretory system.
10 Cerebral and ventral ganglia are connected by circumoesophageal connectives.

Larval form

The young are known as larvae but are similar in form to the adults.

Metamorphosis

No metamorphosis occurs.

Adult body form e.g. *Sagitta, Eukrohnia, Spadella.*

The chaetognaths are known colloquially as the arrow-worms. The body pattern is illustrated in Fig. 29.1A,B. The skeleton is a hydrostatic one.

Feeding

Chaetognaths are all carnivorous, being significant predators of other planktonic organisms such as copepods. The animal moves forward very rapidly over a short distance to grasp the prey with the massive spines of the anterior end (Fig. 29.2). The alimentary canal is a straight simple tube.

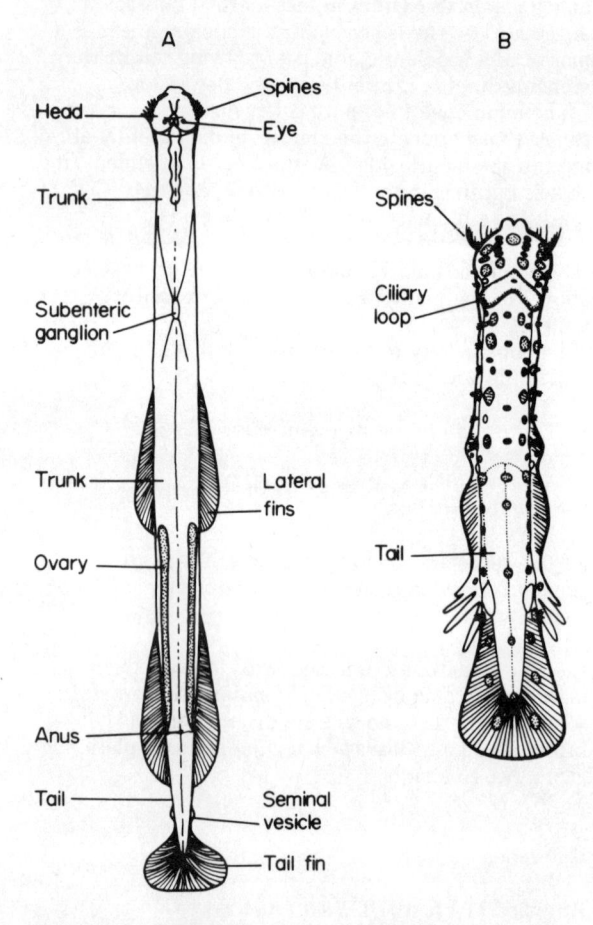

Fig. 29.1 A. *Sagitta elegans*, ventral view showing major features of chaetognath organization (after Ritter—Zahony). B. *Spadella schizoptera*, dorsal view (after Mawson).

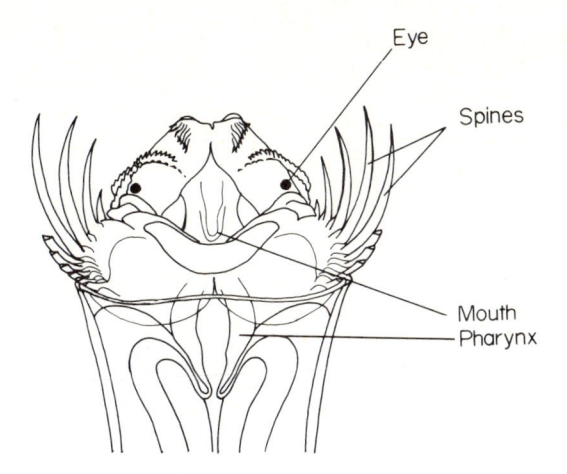

Fig. 29.2 *Sagitta elegans*, anterior region showing spines, eyes and mouth (after Ritter-Zahony).

Osmoregulation/excretion

There are no special organs.

Movement

a The arrow-worms may float or swim. Swimming can be very quick in the form of rapid dashes.

b The lateral fins on either side of the body aid floating by increasing the surface area. They do not appear to play a part in movement.

c Movement is brought about by the contraction of the longitudinal muscle (there being no circular muscle).

Co-ordination

a The nervous system is relatively simple with a massive concentration of tissue anteriorly and a number of paired nerves arising from the subenteric ganglion. These serve the various sensory structures of the body, and presumably the musculature. It is likely that giant fibres are present in view of the great rapidity of the forward movements.

Sense organs present are eyes (based on a ciliary structure), stiff cilia (mechanoreceptors) and chemoreceptors.

b Hormones are unknown.

Respiration

No special adaptations exist.

Circulation/coelom

The coelom arises from outgrowths of the archenteron and hence is enterocoelic in origin. The head contains a single coelomic space separated by a septum from the two spaces of the trunk. The tail is occupied by a single coelomic space. The coelomic fluid acts as a circulatory system.

Reproduction

a Chaetognaths are hermaphrodite with the ovaries anterior to the trunk:tail septum, and the testes posterior to the trunk:tail septum.

b The sperm are enclosed in a spermatophore and released to the outside by bursting of a seminal vesicle. Self-fertilization occurs in *Sagitta*, sperm entering the gonoduct, which opens in front of the septum, and migrating up the oviduct. *Spadella* exhibits cross-fertilization.

Indicator species

Species of chaetognaths are taken as indicative of various oceanic and coastal water masses. Thus in British waters *Sagitta elegans* is typical of phosphate-rich water stemming from the Atlantic and containing some substance that actively encourages larval growth. *S. setosa* occurs in phosphate-poor water, without the larval-promoting substance, and is more typical of coastal waters. The two species are readily identified by the shape and position of the seminal vesicle.

Reference

Ghirardelli E. 1968. Some Aspects of the Biology of the Chaetognaths. In *Advances in Marine Biology* (Ed. F. S. Russel), Vol. 6. Academic Press, New York and London.

30 Phylum ECHINODERMATA

There are about 6,000 living species of echinoderms. In size they range from very small to about a metre across.

Characteristics

1 Exclusively marine.
2 Coelomate, the coelom being enterocoelic in origin.
3 The coelom is arranged in three parts in the embryo and these later develop into the axocoel, the water vascular system and the general coelomic cavity.
4 Symmetry pentamerous.
5 The skeleton is internal and dermal, composed of ossicles of calcium carbonate.
6 There are no excretory organs.
7 There is a unique water vascular system comprising a series of coelomic tubes and superficial extensions that form the podia or tube-feet.
8 Development is radial, and follows the deuterostomatous type in which the anus is derived from the blastopore.
9 Sexes are separate and fertilization is external.
10 The larva is bilaterally symmetrical with a complex metamorphosis.

Larval form

There are several larval types exhibited by the echinoderms (Fig. 30.1) and it has been hypothesised that these may be related to a postulated ancestral type, the dipleurula.

The Crinoidea produce large yolky eggs. Each develops into a barrel-shaped larva the doliolaria that carries 5 bands of cilia around the body. This bears no relationship to the dipleurula.

The other four classes show two major types of larvae:

1 the auricularia, typical of Asteroidea and Holothuroidea;

2 the pluteus, exhibited by Echinoidea and Ophiuroidea.

The auricularia is a small organism with a ciliated band displayed in characteristic fashion around the extremities of the body. A late form of the larva may be termed the bipinnaria and later still the brachiolaria (in Asteroidea) depending upon the development of the arms, or the doliolaria (in Holothuroidea) in which the ciliated band becomes arranged in rings around the larva.

A very late larva, beginning to demonstrate the adult pattern, in crinoids is known as the pentacrinule and in holothuroids as the pentactula.

The pluteus is a somewhat triangular-shaped organism in which the ciliated band is elongated by the growth of arms (which are reinforced by spicules internally).

Metamorphosis

In each case metamorphosis is dramatic and leads to a complete reorganization of the body of the animal. The orientation of the body changes and the terms dorsal and ventral, left and right as applied to the larva are not applicable to the adult (Fig. 30.2A).

In the ideal dipleurula the coelom buds from the archenteron as three pairs of sacs. In an asteroid the development is asymmetrical, with the right side appearing after the left which dominates the process of metamorphosis. The three pairs of sacs soon become modified as in Fig. 30.2B. The water vascular system arises from left sac 2, joined to left sac 1 as the axial organ and axocoel leading to the madreporite. The rear portion of the left sac 2 pushes out small growths from the body wall which determine the main radii of the adult.

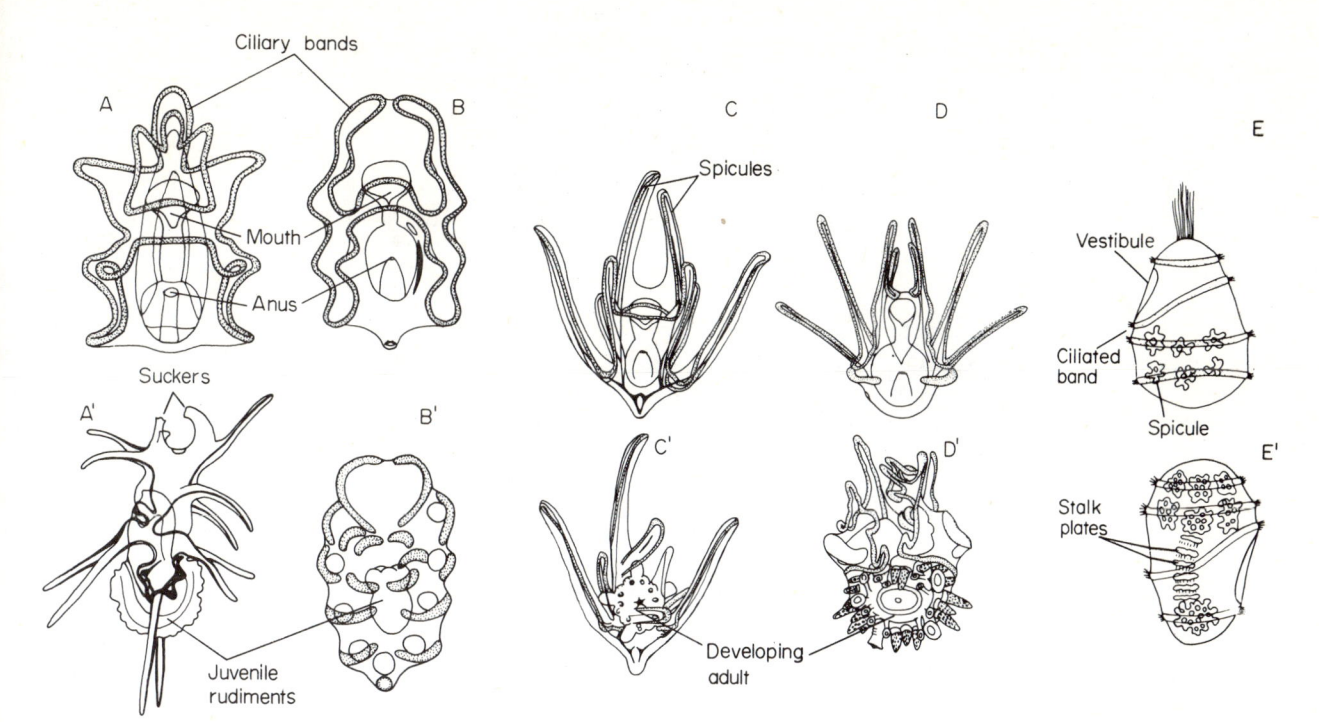

Fig. 30.1 A. Bipinnaria larva of an asteroid (ventral view). This larva has 2 ciliated bands. A[1]. The brachiolaria stage of asteroid development showing suckers for temporary attachment and the developing adult rudiment. B. Auricularia of a holothuroid with one ciliated strip. B[1]. A different form of holothuroid larva (doliolaria) in which the cilia are arranged in bands, later in rings. C. Ophiopluteus of an ophiuroid (ventral view). C[1]. Metamorphosing stage of an ophiopluteus. D. Echinopluteus of a regular echinoid (ventral view). D[1]. Metamorphosing stage of an echinopluteus. E. Doliolaria stage of a crinoid with ciliated bands around the body. E[1]. Late doliolaria, in the inverted stage attained by anterior attachment. Stalk plates appear (A—E[1] after Clark). Not to scale.

Adult body form

The phylum demonstrates one very noticeable and unique characteristic, that of pentamery. Essentially this is a form of organization in which the body is arranged symmetrically in five equal parts around a central core.

This is best displayed by the Ophiuroidea in which a central disc is quite distinct from the long, sinuous arms (Fig. 30.3B). This is less marked in the Asteroidea (Fig. 30.3A). The Crinoidea maintain a pentamerically-organized body form although this is obvious only in the crown (that part of the body at the apex of a long stem). The rays may be subdivided so that 10 or more are found.

Pentamery is less obvious in the remaining classes, the Echinoidea and Holothuroidea. In the globular or disc-shaped echinoids the arrangement of the ambulacra and the jaws of Aristotle's lantern is pentamerous. Aristotle's lantern is a stout, toothed structure operated by strong muscles during feeding. The 5 ambulacra radiate from a central point, the angles between all adjacent pairs being equal. In the Holothuroidea even this feature is obscured since the ambulacra are found in two groups: the bivium (2 ambulacra), and the trivium (3 ambulacra) adjacent to the substratum.

Orientation

It is generally agreed that the primitive echinoderms were sessile, attached by a stalk. The mouth and

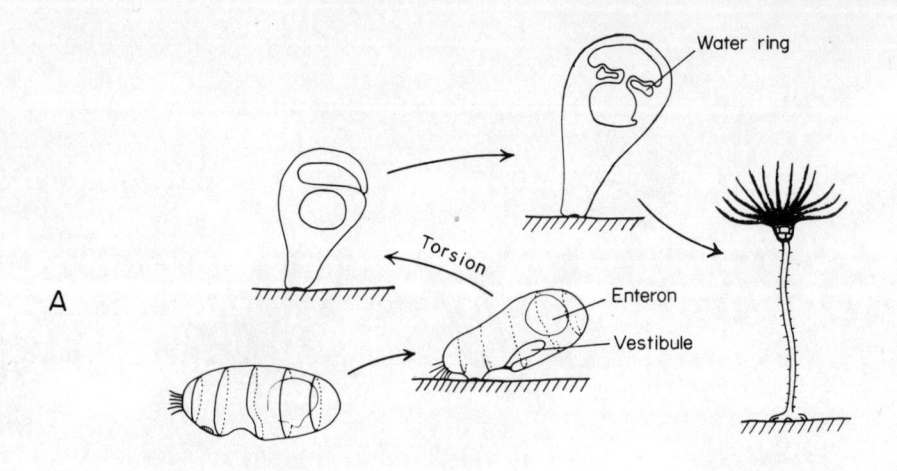

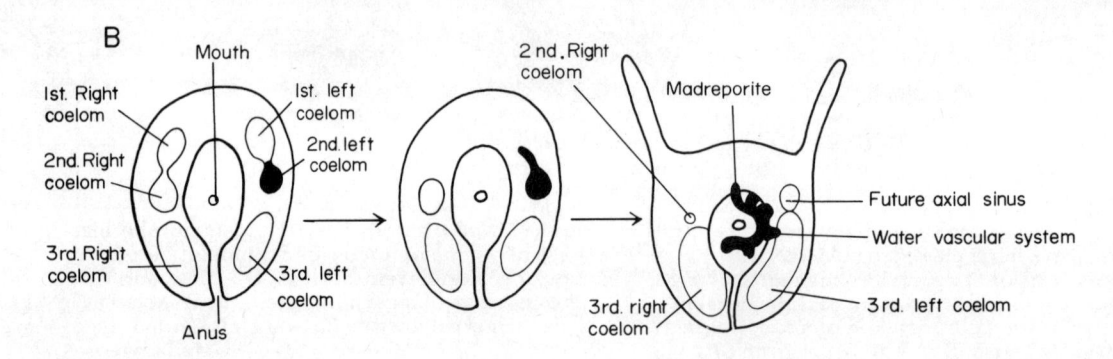

Fig. 30.2 A. Larval development of an attached crinoid. The barrel-like doliolaria settles on the anterior side, attaches by an adhesive plate, and then undergoes torsion to bring the vestibule uppermost. This breaks open to the environment and projections of the water-vascular system protrude into it to form the arms (after Nichols). B. The development of the coelom in an asteroid, indicating the fate of the initially-formed 3 pairs of coelomic sacs (after Borradaile, Eastham, Potts & Saunders).

anus were both directed upwards. Only the crinoids amongst modern echinoderms maintain this posture. The other classes show different orientations relative to the substratum. The surface bearing the mouth is the oral side, and that opposing it is the aboral surface. The surface bearing the tube-feet is the ambulacral surface. These correspond with left (oral, ambulacral) and right (aboral) sides of the larva.

The extant echinoderms fall into three major groups:

1 Sub-phylum CRINOZOA: includes the sessile echinoderms; usually with a stalk and an anus opening close to the mouth.

Class CRINOIDEA e.g. *Antedon, Rhizocrinus*. Primitively possess a stalk by which they are attached to the substrate. Attachment brings the oral surface to face uppermost; there are branched arms bearing tube-feet that lack suckers; madreporite, spines and pedicellariae are lacking. Comatulid crinoids lose the stalk after the pentacrinule larval stage and the animal is free-living.

170   Phylum ECHINODERMATA

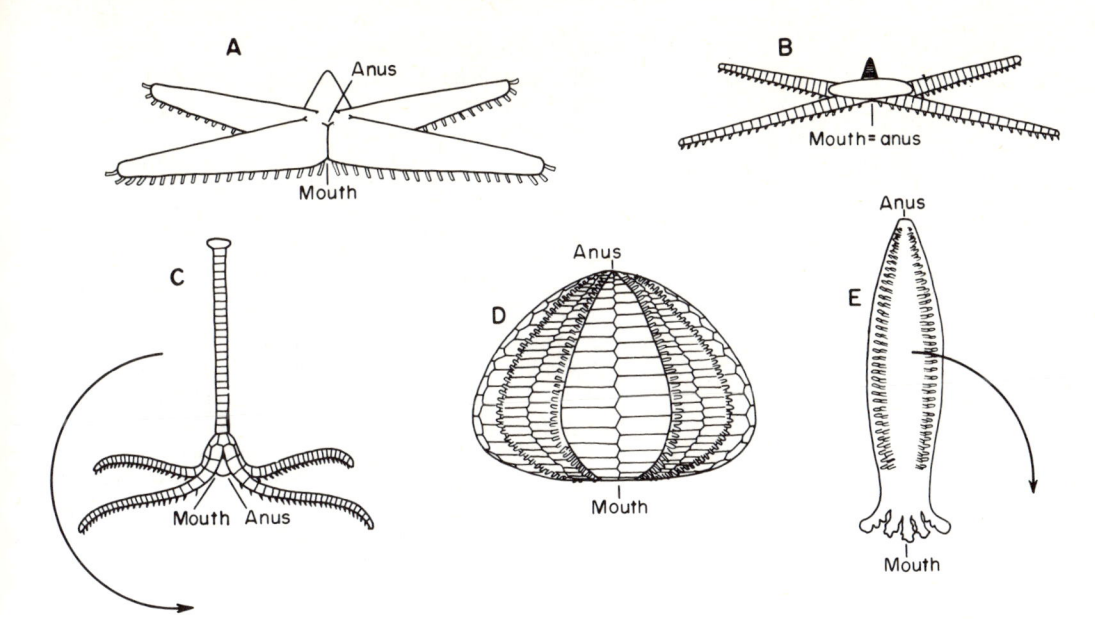

Fig. 30.3 Diagrammatic representation of echinoderm types, viewed inter-radially. The oral side is downward. Arrows indicate normal posture in life of crinoid and holothuroid. A. Asteroid. B. Ophiuroid. C. Crinoid. D. Echinoid. E. Holothuroid (A—E after Clark).

2 Sub-phylum ASTEROZOA: star-shaped free-living echinoderms without a stalk.

Class ASTEROIDEA e.g. *Luidia, Crossaster, Astropecten, Asterias.*
Star-shaped, although the clarity of divisions between the arms may be blurred by the development of webs; arms not sharply delineated from the central disc contain extensions of the alimentary canal (caeca); arms bear numerous tube-feet that have suckers except in burrowing forms; madreporite aboral in position, as is the anus; ambulacral grooves open; pedicellariae sometimes present.

Class OPHIUROIDEA e.g. *Gorgonocephalus, Ophiothrix, Ophiocomina.*
Basket-stars and brittle-stars (so called because of their great fragility). They are star-shaped, with well-defined arms sharply marked off from the central disc which do not contain branches of the gut; there is no anus; the madreporite lies on the oral surface; each ambulacral groove is covered by a row of ossicles but the nervous system remains in contact with the external environment via the epineural canal; tube-feet unsuckered; pedicellariae lacking.

3 Sub-phylum ECHINOZOA: free-living, spherical or sausage-shaped echinoderms without a stalk.

Class ECHINOIDEA e.g. *Cidaris, Echinus, Echinocyamus, Echinocardium.*
Globular, oval or disc-shaped without arms; ambulacral grooves covered and radial nerve cords hence protected from the environment; tube-feet suckered; aboral madreporite and anus present; the whole animal is covered with numerous spines, knobs and pedicellariae.

Class HOLOTHUROIDEA e.g. *Cucumaria, Holothuria, Synapta.*
The body is elongated in the oral-aboral axis and forms a cylinder without skeletal rigidity. The muscular body wall is provided with small ossicles (Fig. 30.4) but there are no other hard parts such as spines or pedicellariae; mouth and anus are terminal at opposite ends of the body; the mouth is

Phylum ECHINODERMATA

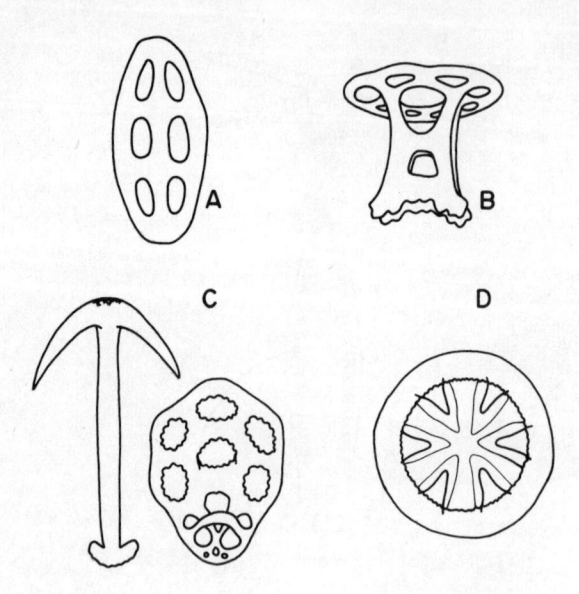

Fig. 30.4 Some spicule types of holothuroids. A. Smooth button. B. Table. C. Plate and anchor. D. Wheel (A—D after Clark).

surrounded by extensible tentacles that are formed from modified podia; the tube-feet of the body (where present) are often short and broad and are suckered; there is usually no external madreporite; radial nerve cord of the ambulacral groove is covered.

Two forms of skeleton are demonstrated by the echinoderms.

1 The dermal skeleton
This is constructed of hard, rigid plates, having different shapes and thicknesses according to the site of formation. The plates may be massive and tightly linked to one another as in echinoids, or sculptured and independent as in the holothuroid body wall. The individual ossicles of the skeleton are made of calcite crystals, which are constructed in a characteristic reticulate pattern with large fenestrae or holes piercing the bulk. The separate ossicles are held together by collagen fibres that loop through the channels, and the remainder of the holes are filled with other living tissue. Spicules of this kind form the spines, pedicellariae, test plates and sucker discs.

Phylum ECHINODERMATA

2 The hydrostatic skeleton
The dermal skeleton in most echinoderms is a fairly rigid one, and hence does not allow much flexibility. Some flexibility is, however, incorporated into the system by the development of the water-vascular system which acts as a hydrostatic skeleton. The basic plan is of a peripharyngeal tubular ring connected to radial tubes that extend, one along each arm. Along each radial canal is arranged a series of podia or tube-feet, which are small (but usually extensible), blind sacs protruding into the environment. Extension is brought about by transference of the fluid content of an ampulla into the podium (Fig. 30.5). These podia provide motive power in asteroids and echinoids, feeding devices in crinoids and ophiuroids, and tentacular organs in holothuroids.

The general body cavity, which is fluid-filled, may also be considered as part of the hydrostatic skeleton in holothuroids which are otherwise ill-supplied with skeletal elements.

Feeding

a Echinoderms include carnivores, grazers, suspension or filter-feeders and scavenging detrital feeders among their numbers.

b Crinoids are attached (even comatulids, though not permanently) and lie with the arms forming a bowl with the mouth at the centre. This acts as a collecting surface, with the podia all active in catching small, drifting particles in the vicinity (Fig. 30.6). These are trapped on ciliary strings that are then retracted into a food groove and passed along to the mouth by the collective activity of three different sizes of podia.

Primitive asteroids were probably ciliary and detrital feeders, as a few modern species (*Porania*, *Henricia*) are, but most present-day representatives are predators, especially on bivalves, but also on fish, crustaceans, polychaetes and corals. The stomach is a remarkable organ that can be intruded between the barely-gaping shell valves of bivalve molluscs, where digestive enzymes are released to begin extracellular digestion.

Echinoids are mostly grazing feeders, possibly omnivorous, scraping at algae and other organic material on rock surfaces, e.g. *Echinus*.

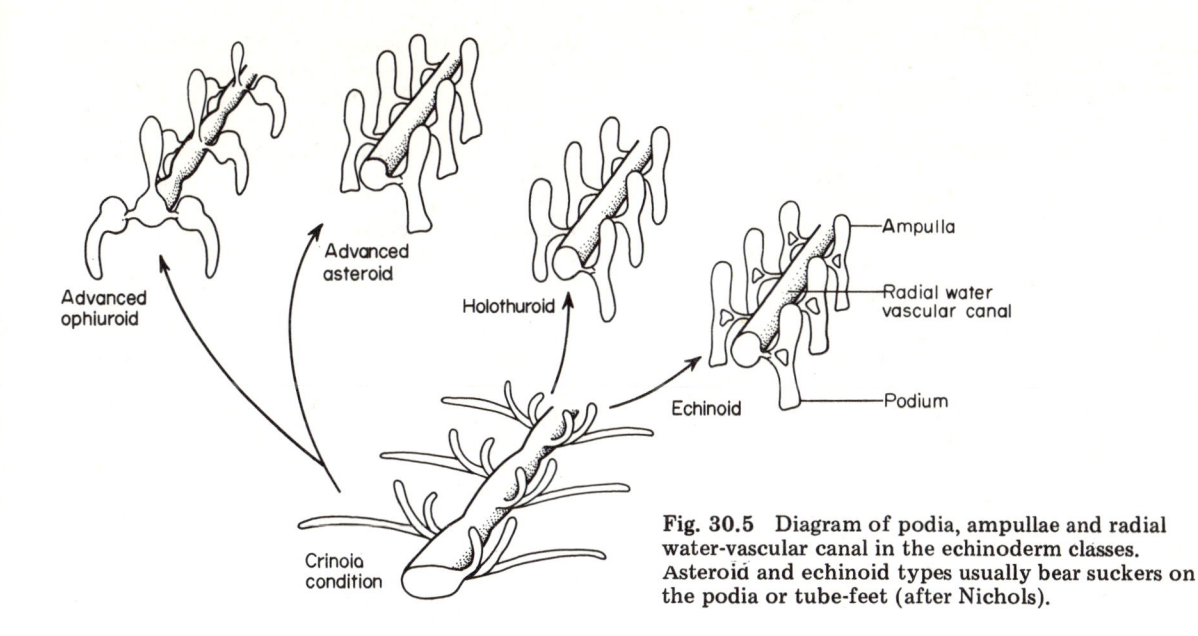

Fig. 30.5 Diagram of podia, ampullae and radial water-vascular canal in the echinoderm classes. Asteroid and echinoid types usually bear suckers on the podia or tube-feet (after Nichols).

Alternatively they may be microphagic particle feeders, utilizing minute organisms encountered during burrowing, or may use specially extensible tube-feet that adhere to particles outside the burrow entrance and retract to bring them back to the mouth, e.g. *Echinocardium*.

Ophiuroids may be ciliary feeders, picking up small organisms on mucus strings. In burrowing forms the arms protrude from the burrow to catch the prey. Some are carnivorous, such as *Gorgonocephalus* with its many branched arms, that catches shrimps or other crustaceans by lassooing them with the mobile arms.

Holothuroids are detrital feeders, sieving the water in front of the frilly tentacles surrounding the mouth, or scooping up the substratum and pushing it into the mouth. Some simply burrow through mud or sand, ingesting it as they go and digesting the organic content.

A number of asteroids and echinoids have the capacity to absorb dissolved organic substances directly from sea water.

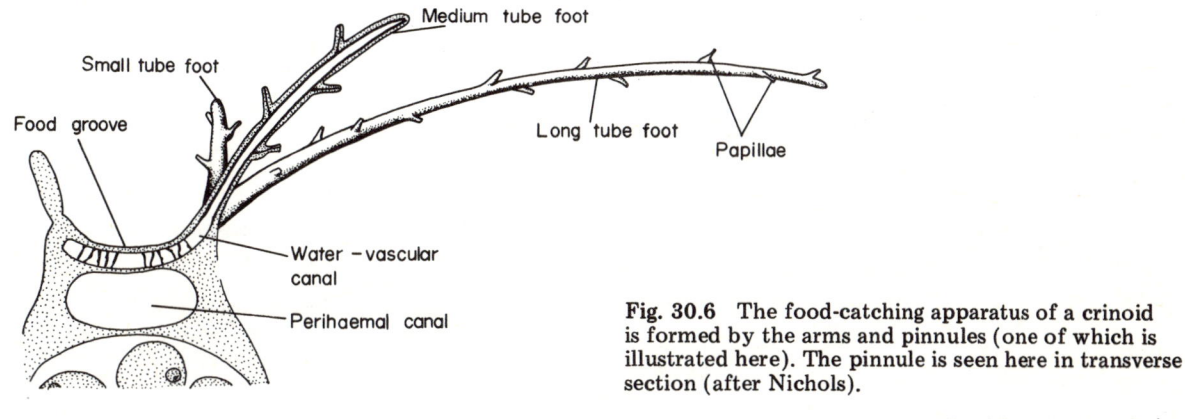

Fig. 30.6 The food-catching apparatus of a crinoid is formed by the arms and pinnules (one of which is illustrated here). The pinnule is seen here in transverse section (after Nichols).

Phylum ECHINODERMATA

Osmoregulation/excretion

a There are no excretory organs.

b Most echinoderms are marine, but a few are euryhaline, appearing in brackish water areas such as the Baltic and Black Seas. They are unable to colonise estuaries. The group as a whole is one of osmotic conformers, there being no ability to control their internal osmotic pressure. There is limited ionic regulation, especially of calcium and potassium, but on the whole the integument is readily permeable to ions in both directions.

Excretion is not carried out by specialized organs, but apparently by coelomocytes that take up foreign material intruded into the coelom and then migrate through the body wall to the exterior, or are released through the papulae (thin-walled respiratory projections) that occur over the body surface.

Respiration

a A variety of structures are involved in respiratory exchange. In animals with fairly rigid skeletons any portion that is thin-walled and where internal fluids come close to the surface may function as an exchange site. In echinoderms this is true for papulae (asteroids), small protrusions of ciliated, external epithelium covering a layer of connective tissue that overlays the coelomic epithelium; for gills (some echinoids) which are located near the mouth and are composed of thin-walled sacs; for tube-feet (in most species) that bring water-vascular fluid close to the surface; and for respiratory trees (holothuroids) which are diverticula of the hindgut, also thin-walled, into which water is pumped rhythmically several times a minute.

b Gaseous exchange can occur across all of these essentially similar structures where internal and external environments are separated by the narrowest of gaps and the thinnest of cellular layers.

Circulation/coelom

Three systems must be considered under this heading: (i) the water-vascular system, (ii) the general body cavity, and (iii) the haemal system. The perihaemal system surrounds the haemal system and is composed of small coelomic sinuses.

The larval coelom (tripartite) gives rise to the axocoel, the water-vascular system, and the general body cavity. The axocoel is closely applied to the stone canal (a calcified organ) and runs from the aboral ring sinus to the oral ring around the mouth. It also is connected with the haemal system and the genital strands.

i The water-vascular system is the unique organ system of the echinoderms. It is laid out in pentameric array, with an oral ring and 5 radial canals bearing the tube-feet (Fig. 30.7). The single stone canal leads to the aboral surface, and opens into a small ampulla which connects through many pore canals to the external surface. These pores perforate a plate, the madreporite, and water from the environment can pass across it into or out of the body. If all tube-feet contract at once a very small amount of fluid passes out. Most of the volume is redistributed amongst the internal tubular spaces. The madreporite may be a pressure equalization device for efficient working.

Each tube-foot possesses an ampulla (small in ophiuroids) and has valves to prevent water-flow in certain directions. The distal end of the podium is suckered for adhesion in many asteroids and echinoids.

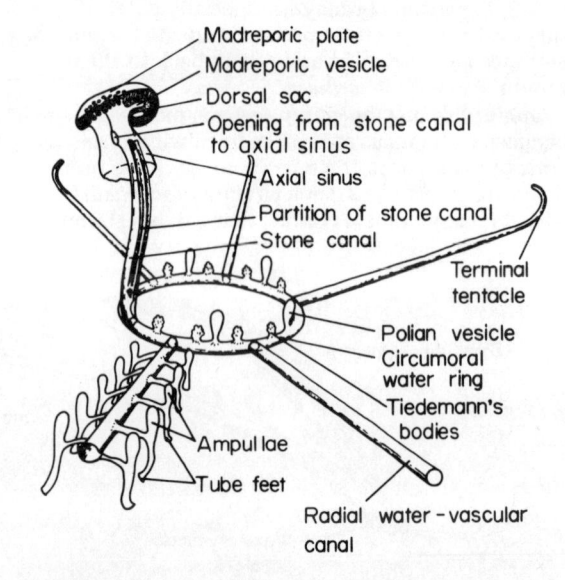

Fig. 30.7 The asteroid water-vascular system, indicating the pentameric plan, the podial series, the position of stone canal, axial sinus and madreporite (after Nichols).

ii The general perivisceral cavity is the fluid-filled space in which all the internal organs of the body are located. It is large in all cases relative to the size of the body, especially so in echinoids and holothuroids. Cilia lining the cavity maintain a circulation of fluid continuously.

iii The haemal system is composed of tissue strands that are located in close proximity to the radial water-vascular canals. The strands are spongy, without distinct walls and epithelial linings. Parts of the system are reported to contract rhythmically but there is no definitive evidence that circulation takes place, nor even that there is a continuous space in which material can flow from one part of the body to another. There is some evidence that coelomocytes, of which there are many kinds in echinoderms, are produced in the tissue, and may be important in defence mechanisms against disease.

Movement

a Several styles of movement may be observed in echinoderms. Attached crinoids sway to and fro and can move the arms; free-living comatulid crinoids can crawl about, and may also swim when disturbed. Asteroids roam on the tube-feet, usually slowly, and show some righting reactions when inverted; some such as *Astropecten* burrow. Holothuroids also burrow in mud and sand (Fig. 30.8C) and some swim. Ophiuroids may live on the surface of mud or sand but can burrow (*Amphiura*, Fig. 30.8A). Echinoids of regular type move by the use of spines and tube-feet; irregular urchins usually live in burrows (Fig. 30.8B).

b Movement is accomplished in a number of ways. Swimming occurs in crinoids as the result of the lashing of the arms; in *Antedon* 5 alternating arms out of 10 are raised and lowered rapidly, followed by the remaining group of 5. In holothuroids such as some elasipods and *Pelagothuria* the body form is modified to assist in flotation and propulsion, presumably by muscular means. Burrowing is achieved by tube-feet moving the substrate away to allow the body to sink into the sand (*Astropecten*, *Amphiura*), by the use of spines (*Echinocardium*) or by muscular contraction of the body wall (as in Holothuroidea). Some species can bore into hard substrates such as wood, rock and even steel (*Strongylocentrotus*).

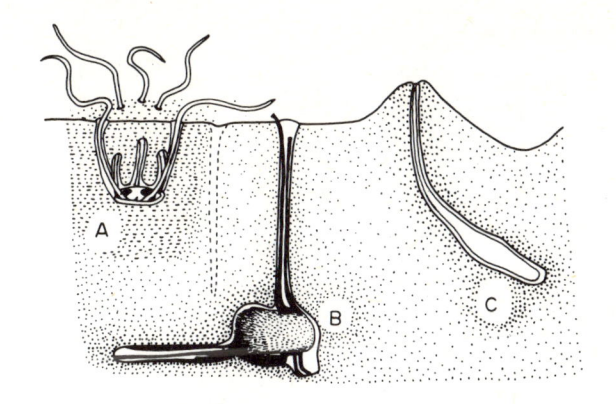

Fig. 30.8 Burrowing is common amongst echinoderms and representatives of each group except crinoids exhibit this mode of life. In this diagram A. *Amphiura*, an ophiuroid, B. *Echinocardium*, an echinoid, C. *Paracaudina*, a holothuroid, are shown in typical attitudes (A—C after Clark).

The familiar slow progression of starfish (*Asterias*) and sea urchins (*Echinus*) is due to the combined and co-ordinated activity of the podia (Fig. 30.9). This can take place in almost any direction with any arm leading, although there is often one arm that leads (dominant radius) more frequently than the others.

c Both striated and non-striated muscles have been described.

Co-ordination

Three types of co-ordination have to be considered amongst echinoderms:

1 the peripheral events of the body surface, especially spine and pedicellarial movement, accomplished via the superficial nerve plexus;

2 the co-ordination and control of the tube-feet, producing directed movement;

3 the synchronization of movement of the five arms of the intact animal, by musculature of the body.

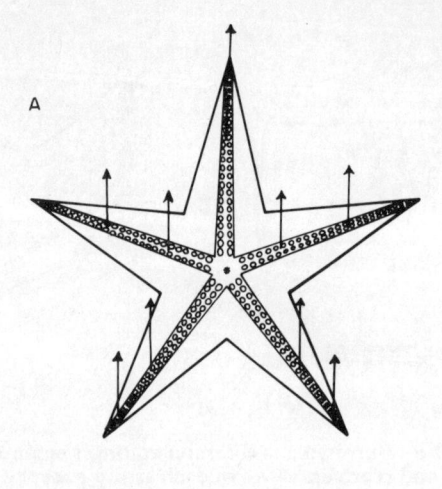

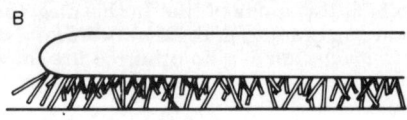

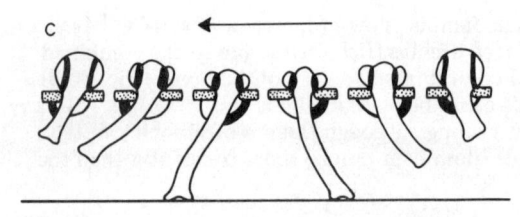

Fig. 30.9 The function of asteroid tube-feet. A. Indicates the directions relative to the radius in which tube-feet show synchronized activity to allow co-ordinated movements. B. Lateral view of an active starfish arm, showing stepping and retraction of the podia. C. Six podia at various stages of the step-cycle showing contraction-relaxation phases of musculature (A—C after Clark).

These three types are not necessarily of equal importance in each group; thus in Holothuroidea body-wall musculature may be more significant than the tube-feet.

a i The neurones are small; sensory cells are bipolar and the internuncial pathways take the form of strands and nets within the epithelia of skin and gut.

176   Phylum ECHINODERMATA

Through-conduction pathways are mainly located within the radial cords of the arms.

The skin plexus is well developed in asteroids, echinoids and holothuroids, whilst in ophiuroids and crinoids it is found only in the walls of the tube-feet, the remainder being obliterated by the ossicles of the body wall. This is the ectoneural system. The arrangement of fibres within this system may differ from class to class. In Asteroidea a non-orientated net of short fibres would explain the radiating and decremental spread of excitation to spines and pedicellariae following surface excitation, whilst the responses of spines in echinoids suggest that excitation may be from spine to spine along straight lines radiating from each receptor site. The fibres of the ectoneural system are organized in the deepest layers into linearly-directed bundles reaching the mid-ambulacral line, where they join the radial cords. Information passes in both directions along the cord, and may pass around the circumoral ring.

A deeper-lying system of nerve fibres exists, the hyponeural system, but is not present in crinoids or ophiuroids and is restricted in echinoids. It is believed to contain the major motor tracts to the musculature. The motor neurones are located in bilateral, metamerically-grouped areas along the radial cord. Ophiuroids have 'giant' axons. The hyponeural axons extend outward across the perihaemal sinus to the vicinity of the tube-feet ampullae, the ossicles and arm sheath. There they make synaptic contact, at least in the case of tube-foot muscles, with prolongations of the muscle fibres that extend down the ampullary seam. These projections are apparently conducting regions only and contain no contractile apparatus (Fig. 30.10).

Sense organs, as collections of receptor neurones, are sparse in echinoderms. There are many sensory cells in the epithelia of the body wall but they are usually single elements, normally bipolar in shape. Up to 4000 sense cells mm^{-2} have been reported for asteroids. In some locations sense organs are described, e.g. on pedicellarial jaws there are mechanoreceptors with stiff cilia; at the ends of asteroid radii there are 'optic cushions' carrying light-sensitive cells; and some holothuroids possess statocysts near the circumoral ring. Otherwise the generally scattered cells seem to be mechanoreceptors or chemoreceptors.

ii Conduction in the epithelial nerve net is decremental, spreading from the point of

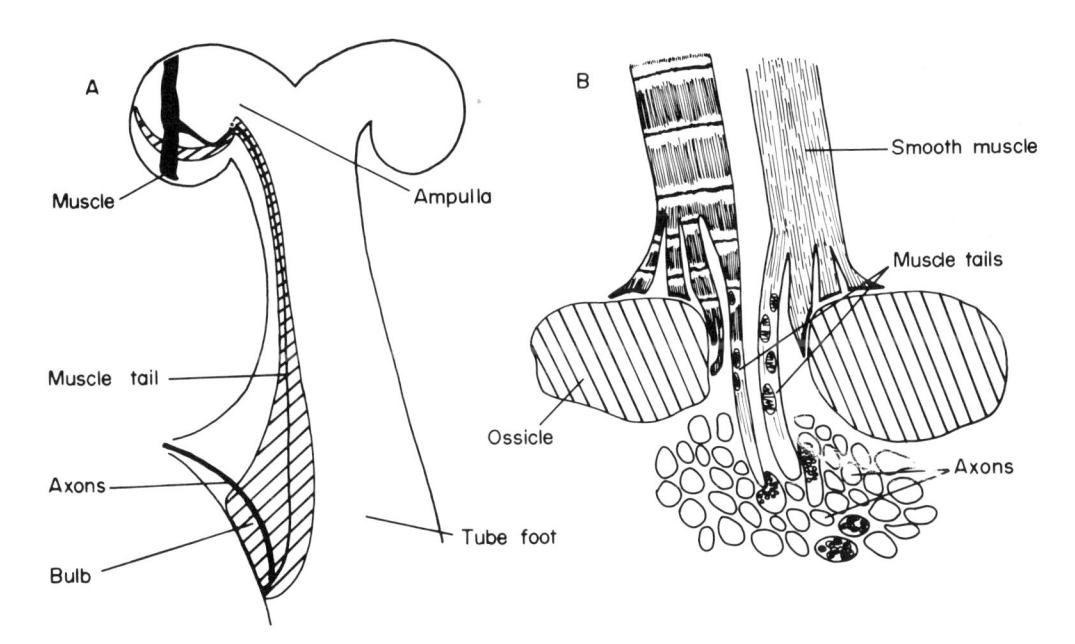

Fig. 30.10 The connection of the nervous system and muscle fibres is achieved via muscle-tails (non-contractile portions of the muscle fibre). A. The organization of muscle, muscle process or tail, and nervous system in a tube-foot. B. Smooth and striated muscle of a pedicellaria. Muscle tails pass between ossicles to a neuropile (A and B after Cobb & Laverack).

stimulation. Excitation may spread along radial lines (not traversing a cut), only travelling in straight lines. The movements of spines and pedicellariae seem governed by such events but local responses of pedicellarial jaws (Fig. 30.11) are completely independent with, effectively, a ganglion located within each of the jaw ossicles where sensory neurones synapse with interneurones and motor elements ('tails' on the muscles).

Control of tube-feet protraction and retraction, necessary for stepping movements, is accomplished via the radial cords. There is through conduction in the cord but also rapid decrement. Changes in pace of stepping and direction pointing appear simultaneously in all podia, suggesting a common pathway governing the separate elements.

Synchrony of movement of all five radii is essential for unidirectional progression. Ophiuroids use arm muscles to cross a substratum and move rapidly; asteroids and echinoids use the tube-feet. It is known that one arm frequently leads, but that the leading arm changes from time to time. The areas from which the pattern is generated seem located in the circumoral ring at the base of each radius (Fig. 30.12). Centres of dominance at the base of each arm are hypothesised, but have not been correlated with anatomical structures.

b Hormonal interactions are not known in echinoderms although the well-known synchrony of reproductive maturity and spawning is a function of circulating chemicals, including pheromones. Hormones controlling osmotic pressure and metabolism are indicated.

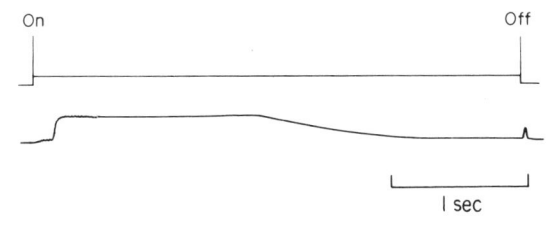

Fig. 30.11 The closing response of a tridentate pedicellaria to a mechanical stimulus is very rapid, closure is maintained for a short while, and then the jaws open again slowly (after Campbell & Laverack).

Phylum ECHINODERMATA

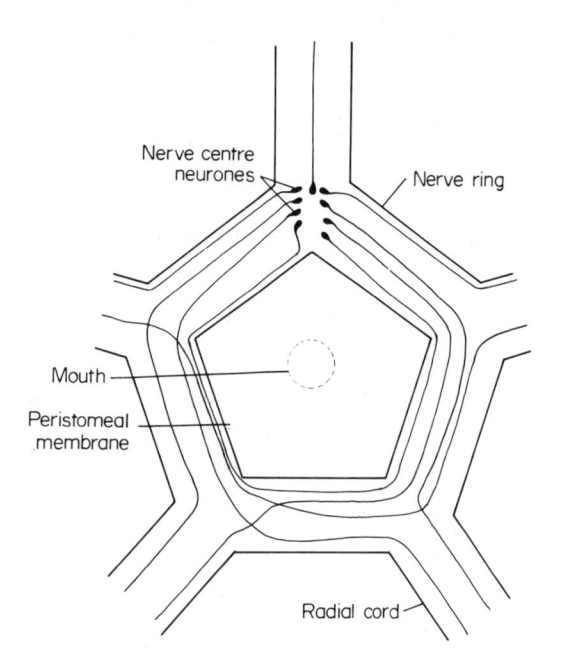

Fig. 30.12 The co-ordination of tube-feet activities is postulated to be due to the activity of centres lying at the base of each radial arm, with connections via axons to all other arms. Information passes both ways around the circumoral ring, and tube-feet are able to point in more than the five primary directions. This figure indicates only one dominant centre and radiating pathways. All radii possess an equivalent plan. The radial nerve connects to axons, and muscle tails of effector organs (after Smith, with information from Kerkut).

Reproduction

a   The sexes are separate, except in a few unusual cases, e.g. *Asterina*. In crinoids the gonads are numerous and are placed along the arms. Asteroids, echinoids and ophiuroids all have five gonads in the main part of the body. Each opens externally by a single pore. Holothuroids have a single gonad.

b   Fertilization is external. Echinoderms are often gregarious. Populations of echinoderms may shed gametes synchronously, co-ordination being brought about by chemical means between individuals although the processes of maturation have also to coincide in time. Development is total, indeterminate and radial. A few members of each class brood the embryos as in the genital bursae of ophiuroids, or the stomach of some starfish (*Leptasterias*). In some starfish the embryos are physically attached to the parent. There are other methods of brood protection.

Regeneration

As a phylum echinoderms show a remarkable capacity for regeneration. A few robust species, not normally much damaged, are capable only of wound healing and repair. The majority, however, are able to replace one or more lost arms, and some can restore parts of the disc. *Linckia* may regrow a disc and all arms from a piece of one arm. A form of asexual reproduction may result since fragmentation and restoration occurs frequently. Autotomy is known in a number of asteroids. It also occurs in holothuroids such as *Leptosynapta inhaerens* and *Cucumaria planci*. Crinoids and ophiuroids both exhibit good powers of regeneration but echinoids seem limited in their capacity.

References

Boolootian R. 1966. *Physiology of Echinodermata*. Wiley, London.
Clark A.M. 1962. *Starfishes and their relations*. British Museum, Natural History, London.
Nichols D. 1969. *Echinoderms* (4th edition). Hutchinson University Library, London.

31 Phylum UROCHORDATA

About 2,000 small to medium-size species, individuals up to several cm, colonies may reach metres in length or breadth.

Characteristics

1. All marine.
2. Adult has no coelom, shows no segmentation and possesses no bony tissue.
3. There is a test of tunicin (a substance related to cellulose).
4. There is a dorsal atrium.
5. Notochord is restricted to the tail, and is found only in the larva, except in one class.
6. CNS is removed from the surface and is degenerate in the adult.

Larval form

Ascidiacea Compound ascidians retain the larva within the atrium until it can swim. Larvae are most complex amongst compound, simplest amongst solitary forms. The typical form is the ascidian 'tadpole' (Fig. 31.1). Larval life may last a few hours to several days, the larva showing initially positive phototropism and negative geotropism, and later a decrease in light response and a migration to dark areas.

Thaliacea The larval stage is tailed in the doliolids, but there is no larval stage in salps.

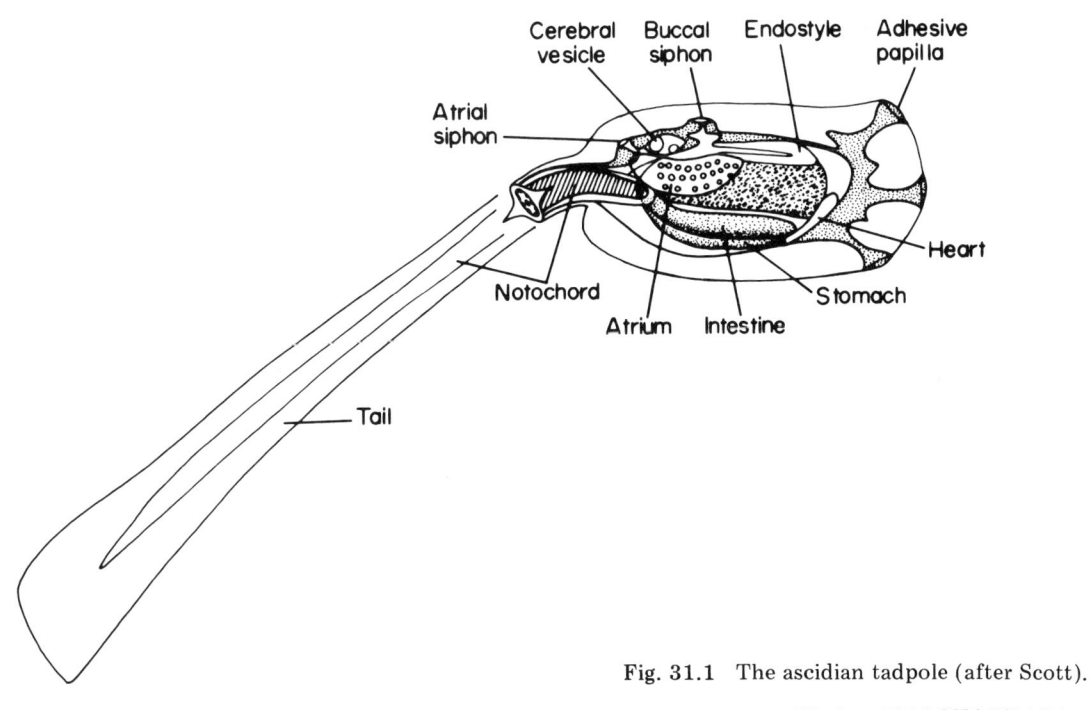

Fig. 31.1 The ascidian tadpole (after Scott).

Metamorphosis

Ascidiacea The reasons for settlement are not properly known, but it probably occurs in response to chemosensory stimuli. The larva attaches to a solid object by adhesive papillae and metamorphosis is complex and dramatic (Fig. 31.2).

Thaliacea The larval stage develops into a nurse zooid (oozooid) whilst still planktonic.

Adult body form

The adult form differs according to class.

1 *Class* LARVACEA (APPENDICULARIA) e.g. *Oikopleura*, *Fritillaria*. These are urochordates in which the adult maintains the larval organization.

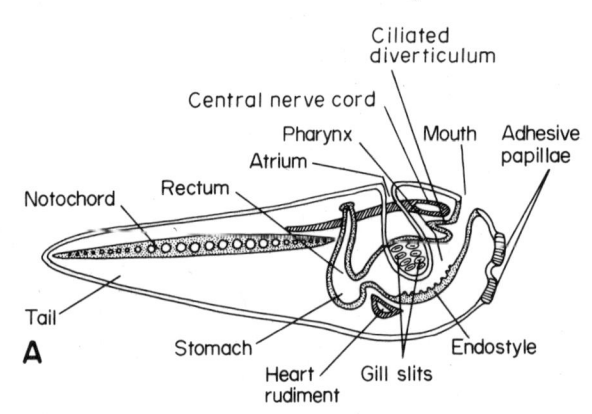

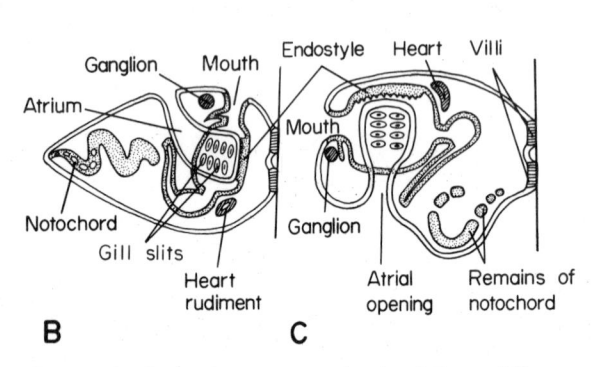

Fig. 31.2 Stages in metamorphosis of the ascidian tadpole. A. Locomotory stage. B. After settlement. C. Rotation of organs (A, B and C after Russell-Hunter).

The individual is planktonic and lives in a secreted house, the test not being of tunicin (Fig. 31.3). The house aids in filtration of food and is not permanent, being withdrawn and resecreted from time to time. Although the appearance is similar to that of the ascidian tadpole there are fundamental differences, namely (a) the presence of gonads, the animal being hermaphrodite and protandrous, (b) the tail is attached at the ventral side, (c) there are only simple gill clefts opening ventro-laterally to the exterior, (d) the brain is compact and there is no cavity in it or the nerve cord, (e) there is an ocellus and a statocyst.

2 *Class* ASCIDIACEA e.g. *Clavelina*, *Molgula*, *Sidnyum*. These are urochordates in which the adult is sedentary, and has no tail. Species may be solitary (*Ciona*) (Fig. 31.4) or compound (*Botryllus*). The test is often massive. The nervous system is degenerate, being reduced to a solid ganglion with radiating nerves. Associated with this is the subneural gland which opens into the pharynx via a ciliated funnel. There is an atrium that opens dorsally, water entering through the mouth (oral siphon), passing across the pharynx into the atrium and then escaping through the atrial siphon. There are several gill clefts which are subdivided by external longitudinal bars. There may be a stolon, simple in form.

3 *Class* THALIACEA e.g. *Salpa*, *Doliolum*, *Pyrosoma*. These are pelagic animals. The adult has no tail, a degenerate nervous system, an atrium that

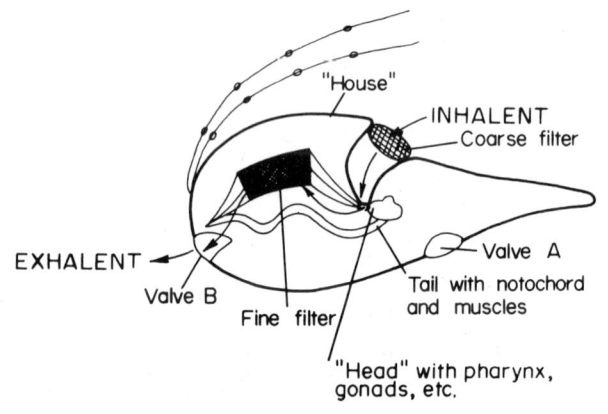

Fig. 31.3 Location of a larvacean in its secreted 'house' (after Russell-Hunter).

180   Phylum UROCHORDATA

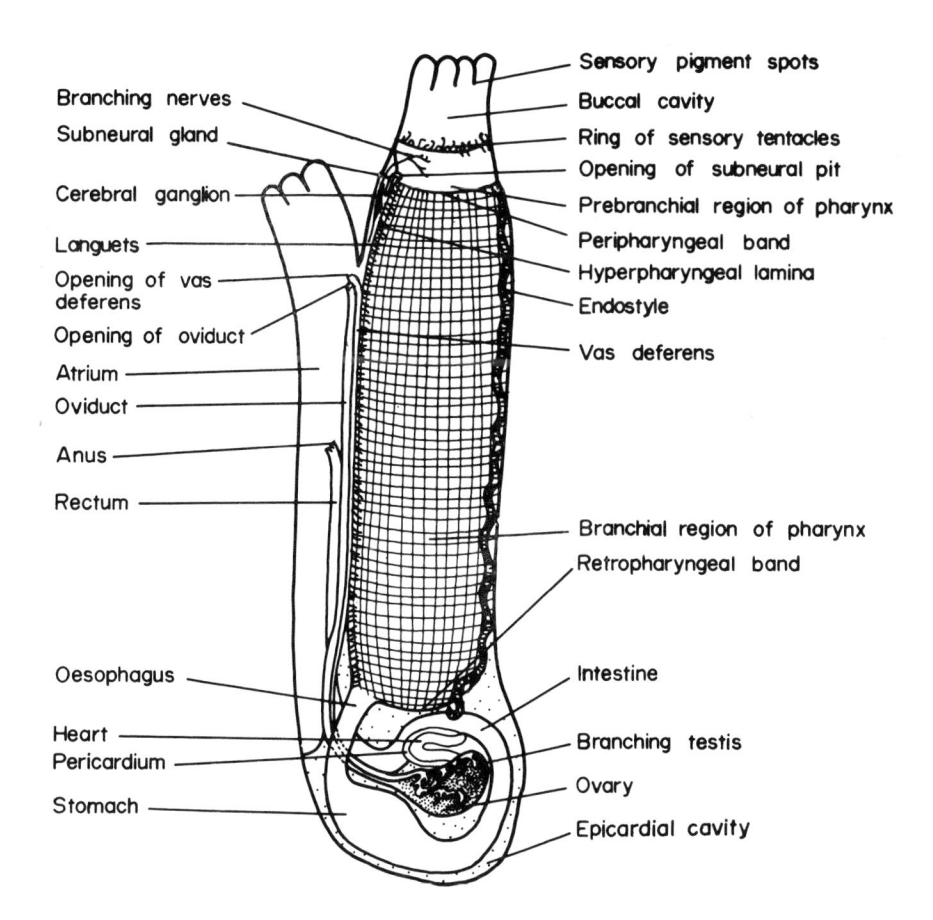

Fig. 31.4 The anatomy of *Ciona* (after Bullough).

opens posteriorly (Fig. 31.5), gill clefts that are not subdivided by gill bars, and a stolon that is complex. There are three orders: Pyrosomidae, Salpidae and Doliolidae, that differ in the degree of muscular development.

Feeding

a In most urochordates phytoplankton is the major component of the food, and filter-feeding the method of obtaining it. There may be some selective feeding with certain species taking particular portions of the plankton. A few unusual ascidians may be scavengers amongst bottom deposits, or even carnivorous. Larvaceans feed on nannoplankton.

b The urochordate pharynx is the feeding organ. Water swept in through the mouth crosses the gill clefts (subdivided in the Ascidiacea to increase surface area) and the solid component is removed by adhesion to mucus produced by the characteristic endostyle in the floor of the pharynx (Fig. 31.6). The mucus string is then passed back to the oesophagus and digested in the gut. Larvacea remove nannoplanktonic organisms by a secreted net that constitutes part of the 'house'. This is periodically retracted into the gut and the whole structure digested.

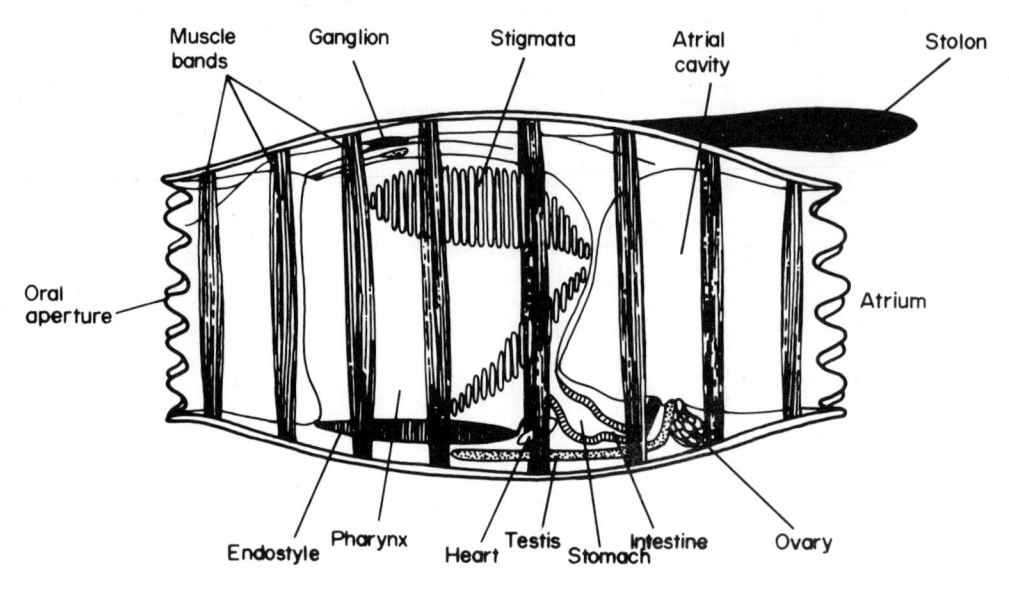

Fig. 31.5 Organization of the body in a thaliacean (after Herdman).

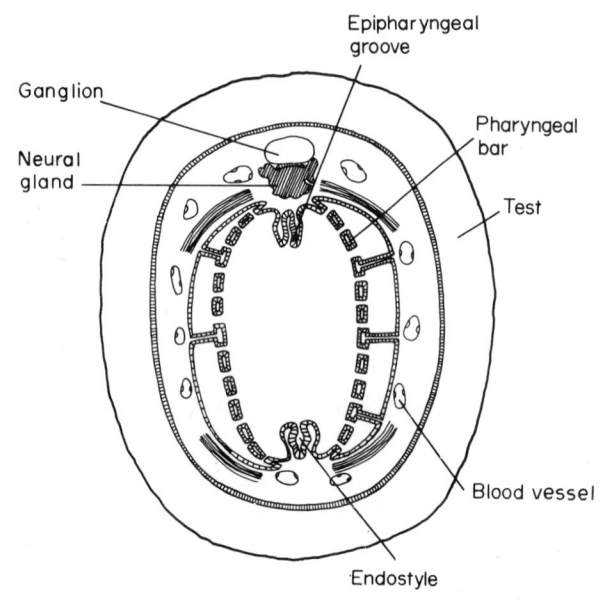

Fig. 31.6 A diagrammatic representation of a transverse section through the body of a urochordate (after Julin).

182   Phylum UROCHORDATA

Osmoregulation/excretion

a No specialized excretory organs exist in the Urochordata.

b Ionic regulation is probably a function of the general body surface. There are, however, some specialized cells within the epicardium (a fluid-filled space around the heart) suspected of involvement. Osmoregulation is probably limited: 6 species only are found in the Baltic Sea and 1 in the Yselmeer.

Movement

Ascidian movement is largely restricted to the opening and closing of apertures (siphons), and the retraction of the body when stimulation is violent. Thaliacean species have muscular bands around the body that provide propulsive power by creating a jet of water through the rearward-pointing atrial siphon. Water movements generally are, at least in part, generated and assisted by the ciliary cover of the gill bars. Larvacea drift when fully enclosed by the 'house', but when, as in plankton collections, the 'house' is destroyed the tail lashes violently. The movement is undirected.

Co-ordination

a The nervous system is a very simple one in adults of Ascidiacea and Thaliacea. In the Larvacea there is some extension along the notochord of the tail, but there is no tubular arrangement of the tissue. Ascidiaceans have a solid ganglion with radiating nerves.

Sense organs in these animals are mainly represented by isolated mechanoreceptor cells found in various places on the body, but especially around the siphons. Photoreceptors seem to be present and chemoreceptors are likely. In all cases there is a greater development of such sensory apparatus in the free-swimming larval stages, which may also possess statocysts.

b There are no definite clues that hormonal regulation is available to any physiological process in urochordates, but circumstantial evidence suggests that at least breeding cycles and spawning are synchronized and may be co-ordinated by hormonal means. The subneural gland has been implicated in hormone production, lying as it does in close proximity to the ganglion and the pharynx.

Respiration

This is probably a function of the pharynx, but there is no substantive evidence as to the mechanisms involved.

Circulation/coelom

a There is a blood system containing a colourless fluid. There is a heart. There is no definitive evidence for the presence of a true coelom, though small sacs near the heart (epicardium) that may be enterocoelic in origin, could be the remnants of a larger coelom.

b The heart propels the blood through the vessels, although the vessels are not walled in many places and should properly be considered as spaces between other tissues. Circulation is not continuous since the heart reverses its direction of beat periodically. The heart has a wall one cell thick, the cells being muscular but lacking T-tubules, and with fibrils limited to one third of the cell. The beat is peristaltic and is governed by pacemakers located at either end of the heart. Each cell is electrically coupled to its neighbours. The pacemakers are affected by changes in partial pressure of carbon dioxide and hydrostatic pressure.

c There are no blood pigments as such, but numerous corpuscles circulate through the body and some of these are highly coloured, green, orange and blue being common. Rare metals such as niobium and vanadium are concentrated by the cells in some species.

Reproduction

a Many urochordates are hermaphroditic.

b *Sexual reproduction* may be temperature dependent and seasonal. Spawning may be synchronized by light stimuli. Fertilization is usually external, but some species retain the eggs and brood them until the larvae are formed.

In Thaliacea there is a complex alternation of generations. In salps, for example, the sexual form produces one egg that develops internally in the parent, nourished by a placenta-like organ whose cells migrate into the embryo. When fully formed this oozooid is asexual and produces buds which can develop sex organs.

In *Salpa*, sexually-produced forms are solitary, the asexual generation is aggregate.

Doliolids have a more extreme cycle in which a larval stage develops into a nurse oozooid that buds to give strings of daughter zooids that remain attached. These are of three types:

i trophozooids, which are sterile, nutritive and respiratory in function,

ii phorozooids which are sterile nurse forms, eventually released,

iii gonozooids, sexual forms nursed and carried by phorozooids until mature.

Stolons are present in compound forms. Budding occurs in some families and fission of colonies in some ascidians.

References

Barrington E.J.W. 1965. *The Biology of Hemichordata and Protochordata.* University Reviews in Biology. Oliver & Boyd, Edinburgh and London.

Millar R.H. 1971. The Biology of Ascidians. In *Advances in Marine Biology* (Ed. F.S. Russel) Vol 9, 1–100. Academic Press, New York and London.

32 Phylum HEMICHORDATA

About 100 species known, up to 2 metres in length.

Characteristics

1 All marine.
2 Vermiform, elongate animals.
3 Coelomate with the coelom divided into three parts corresponding to three externally visible portions of the body: the proboscis, the collar and the trunk.
4 There is no post-anal tail.
5 There are gill slits, but no atrium.
6 Possess a 'stomochord' (in older texts referred to as a notochord, but now agreed to be neither analogous nor homologous with the chordate notochord).
7 The hollow nerve cord is restricted to a short region of the mesosome, otherwise the nerve system is superficial.
8 Larval form is the tornaria.

Larval form

The typical larval form of most hemichordates is the tornaria (Fig. 32.1) which bears some resemblance to the auricularia of echinoderms, though this may be due more to convergence than to any specific relationship. Some species have large eggs that develop directly without a larval stage.

Metamorphosis

The metamorphosis of the tornaria (Fig. 32.2) occurs after the larva has left the plankton and descended to the bottom of the sea.

Adult body form

There are three major portions of the body which reflect the internal organization and the

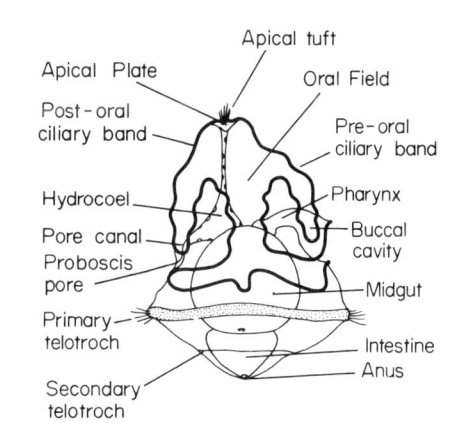

Fig. 32.1 The tornaria larva (after Burdon-Jones).

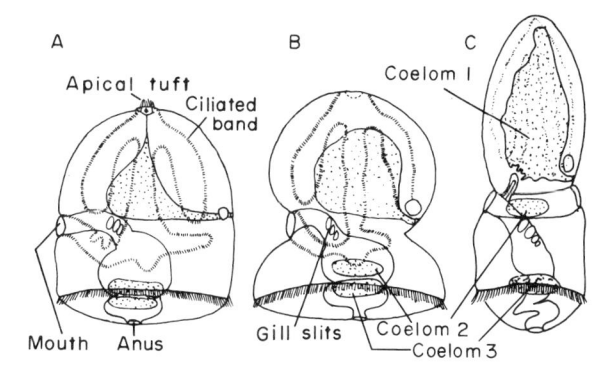

Fig. 32.2 Stages in the metamorphosis of the tornaria larva (after Dawydoff).

arrangement of the coelom. The anterior region is the protosome (proboscis, containing the protocoel), the mid-region the mesosome (collar, enclosing mesocoel) and the posterior region the metasome (trunk, enclosing metacoel) (Fig. 32.3).

The stomochord projects forward from the anterior gut into the proboscis. Alongside it lies the central sinus and the glomerulus. The metasome

Phylum HEMICHORDATA

is punctuated serially by the gill slits that are added to throughout life as the animal grows in length.

The skeleton is hydrostatic; pterobranchs live in secreted tubes.

The living representatives are placed in two classes.

1 ENTEROPNEUSTA e.g. *Balanoglossus*, *Ptychodera*. Free-living, worm-like hemichordates (Fig. 32.4A) that have many gill slits and a straight gut.

2 PTEROBRANCHIATA e.g. *Cephalodiscus*, *Rhabdopleura* (Fig. 32.4B,C). Sessile, tubicolous animals with gill slits reduced or absent, and the gut U-shaped.

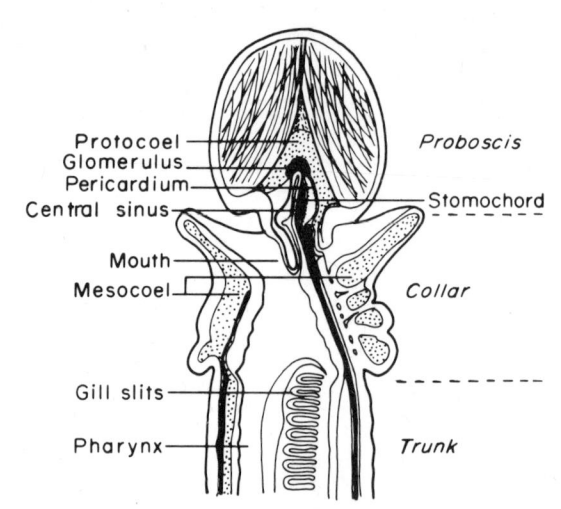

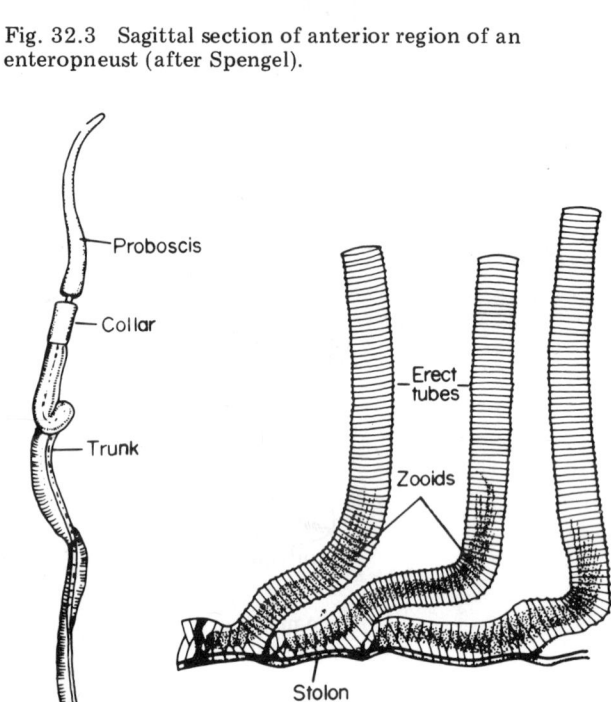

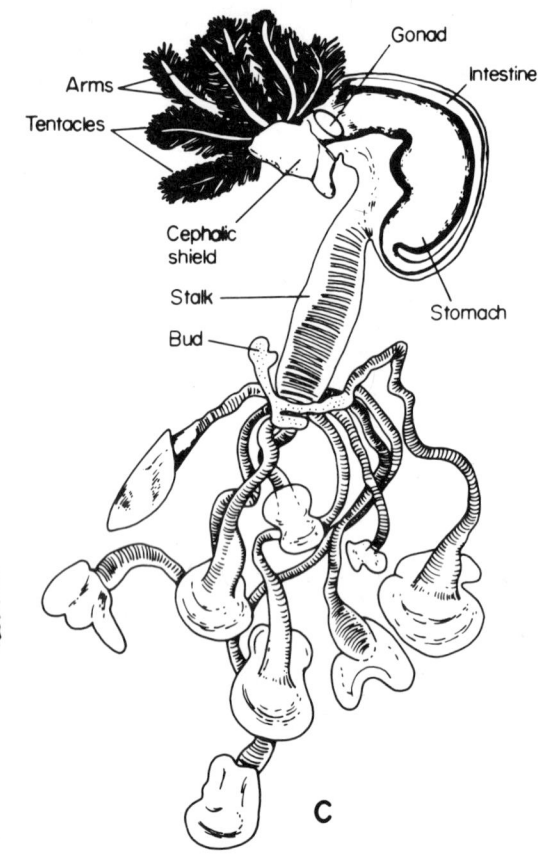

Fig. 32.3 Sagittal section of anterior region of an enteropneust (after Spengel).

Fig. 32.4 A. An adult enteropneust, *Saccoglossus* (after Hyman). B. Tubes and colony of a pterobranch, *Rhabdopleura* (after Schepotieff). C. *Cephalodiscus fumosus* (after John).

The pterobranchiates are colonial, with a number of zooids together in a branched, tubular structure. The individuals remain joined together in *Rhabdopleura* but become separate in *Cephalodiscus*. Each zooid has the typical structure of proboscis, collar and trunk with associated coelomic structures.

The mesosome carries tentacles (2 pairs in *Rhabdopleura*; 5—9 pairs in *Cephalodiscus*). Two gill slits are found in *Cephalodiscus* but none in *Rhabdopleura*.

Feeding

a Feeding is ciliary on microscopic organisms.

b There is a complex arrangement of ciliary fields that collect and process small particles for entry into the mouth (Fig. 32.5). In enteropneusts these cilia are carried on the proboscis and collar; in pterobranchs they are found on the tentacles, which are also sticky. There is anatomical evidence for nervous control of ciliary beat. The gut is elongate with a terminal anus. The walls are perforated by the gill slits. Food is sorted, and rejected material passes rearward in a ventral gutter.

Osmoregulation/excretion

a In Enteropneusta the glomerulus is a complex, tubular structure lying in the proboscis coelom. It is highly vascularized and may be involved in excretory processes. The protocoel connects to the exterior via a coelomoduct.

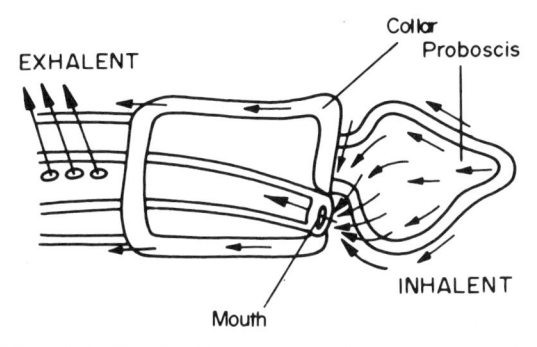

Fig. 32.5 The feeding current of an enteropneust, brought about by surface cilia. Water enters the mouth and passes along the gut, leaving via the exhalent pores or gill slits (after Russell-Hunter).

b Physiology is unknown.

Movement

a Enteropneusta burrow; Pterobranchiata make excursions up and down their tubes.

b Enteropneusta burrow using the proboscis and collar. The principle is that of a hydraulic ram mechanism, with the highly muscular proboscis acting upon the internal coelom in which cilia maintain a constant flow of fluid. Among the pterobranchs, *Rhabdopleura* is attached by a stalk that is very contractile and enables the zooid to retract rapidly into its tube, whilst in *Cephalodiscus* the tail region is adhesive and not continuous with any other structure, thus allowing the zooids to move out of the tubes.

Co-ordination

a *i* The nervous system has affinities with other deuterostomes: a nerve plexus covers the body as an epidermal structure. This is thickened in places to form a CNS and in the mesosome is separated from the epidermis, and rolled to form a dorsal tube. This region contains giant cells and giant fibres. No sense organs have been described apart from scattered cells, although a U-shaped, ciliated depression on the ventral side of the proboscis may be a chemoreceptor; it appears to be heavily innervated in some forms.

ii Giant fibres are concerned with fast conduction along the length of the body. The fibres decussate in the collar. Through conduction along nervous pathways has been indicated by experiment and shown to be involved in such activities as peristaltic movement, burrowing, and spread of luminescence.

b There is no evidence yet on hormonal control mechanisms.

Respiration

a Hemichordates demonstrate pharyngotremy, that is they possess gill slits which are partially divided by tongue-bars (Fig. 32.6). Water enters the mouth as part of the feeding activity, and leaves via the gill

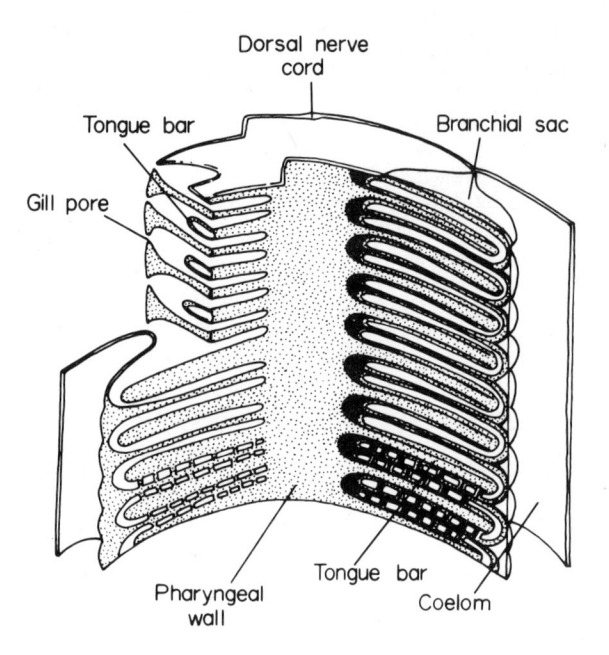

Fig. 32.6 Gill slit structure in the trunk region of an enteropneust (after Delage & Herouard).

slits. The gill apparatus may have evolved initially as a feeding device, and subsequently became modified as a respiratory surface.

b Gaseous exchange presumably occurs across the gill bars into the blood vessels within them. Blood analyses have not yet been carried out to confirm this.

Circulation/coelom

a The coelom is tripartite in the adult and the body divisions reflect this. Both enteropneusts and pterobranchiates demonstrate this phenomenon.
   The blood system consists of a dorsal vessel in which blood moves forward, passing into a central sinus at the base of the proboscis.

b The floor of the muscular fluid-filled sac (sinus) pulsates and drives the blood through the vessels. All blood passes through the glomerulus and thence to the proboscis or backward in the ventral vessel, returning via connective vessels to the dorsal side.

c The blood is colourless.

Reproduction

a The sexes are separate. The gonads are simple sacs anteriorly in the trunk. They open via special pores through which the gametes are shed.

b Fertilization is external.

Reference

Barrington E.J.W. 1965. *The Biology of Hemichordata and Protochordata*. University Reviews in Biology. Oliver & Boyd, Edinburgh and London.

33 Phylum CEPHALOCHORDATA

There are 3 genera, reaching several cm in length.

Characteristics

1. All marine.
2. Elongate and bilaterally flattened.
3. Unpigmented.
4. Pointed at anterior and posterior ends.
5. No recognizable specialized head.
6. Many gill slits in pharyngeal wall. These do not appear externally as they are covered by a fold of skin; the atrium thus formed is a fluid-filled space around the gill slits, opening at the atriopore.
7. Possess a tail, that portion of the body that projects behind the opening of the anus.
8. Muscles arranged in blocks (myotomes) that are easily visible, innervated via 'muscle tails'.
9. There is a notochord composed of muscle fibres.
10. Sense organs apparently located on cirri anteriorly and in a few places along the nerve cord. cord.

Larval form

Amphioxus larvae are similar in form to the adult, although they possess fewer gill slits, have no gonads, and have a club-shaped gland associated with the pharynx that later regresses. They are planktonic and demonstrate vertical migration.

The 'amphioxides' stage is planktonic and represents a 'giant' larva. Adults from such larvae differ slightly from those of small larvae.

Metamorphosis

This process is triggered by contact with the bottom (the adult being bottom-living). Control may be due to the activity of the club-shaped gland, which regresses after the juvenile stage is passed. 'Amphioxides' may result from prolonged planktonic existence.

Adult body form

e.g. *Branchiostoma lanceolatum*, the lancelet or amphioxus; *A symmetron*, with gonads only on the right side of the body.

The typical structure is shown in Fig. 33.1A,B. The presence of a notochord, extending from anterior to posterior tip of the animal, is an important feature.

The notochord is a muscular organ as shown by electron microscopy and physiological evidence. The organization of the muscle is shown in Fig. 33.2. Muscular contraction increases the stiffness of the notochord with a corresponding rise in the internal pressures. Contraction is preceded by a giant fibre action potential.

The notochord functions as a variable stiffening rod, preventing shortening of the body during swimming. It is a kind of hydrostatic skeleton, whose functioning is co-ordinated with the operation of myotomal muscle. Myotomal muscles are serially arranged in a characteristic V-shaped fashion, as in vertebrate segmentation.

There is a dorsal fin and 2 ventro-lateral folds that are fin-like, but not to be confused with such structures in fish. The dorsal ridge is supported by fin ray boxes of connective tissue.

Feeding

a Amphioxus is a microphagic suspension-feeder.

b Water is drawn in under the oral hood, and the cirri and velum are active in sorting large particles. Entry and passage through the pharynx is brought about by the beat of the cilia that cover the many gill bars. An endostyle in the ventral floor produces mucus that entangles food particles, passes up the gill bars and then back to the oesophagus. The water current moves into the atrium and leaves via the atriopore. The alimentary canal is straight, opening at the anus ventrally, although a mid-gut diverticulum pushes forward into the pharyngeal region.

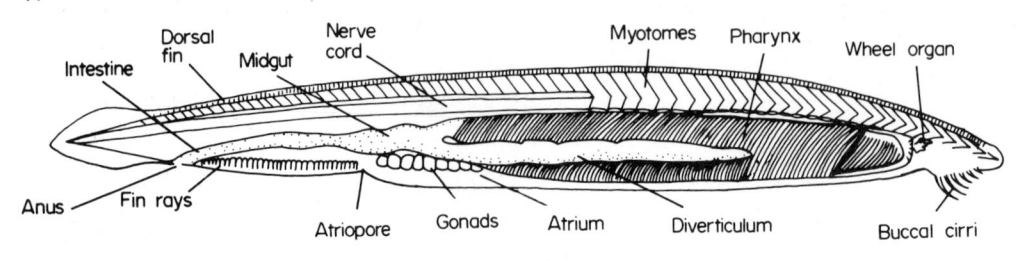

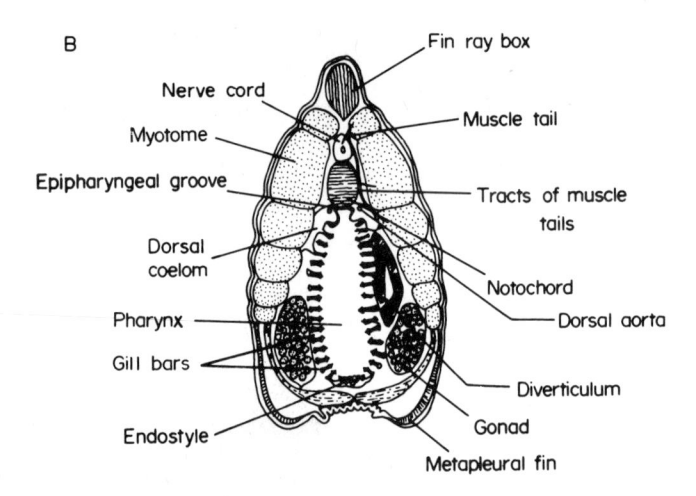

Fig. 33.1 The adult body structure of *Branchiostoma*. A. In transparency. B. Transverse section of the pharyngeal region (A and B after Young).

Movement

a The characteristic attitude of amphioxus when undisturbed is buried in the sand with only the anterior oral hood above the surface. When flushed out of the sand in which it lies, the animal swims quickly with serpentine movements of the body. Burrowing is accomplished head first, rapidly among loosely-packed particles, more slowly where the grains are fine.

b The muscles of the body are arranged in myotomal blocks of striated fibres separated by regions of connective tissue (myocommas). Myotomes are serially repeated in a segmental fashion. Contraction results in a transverse motion of the body, inclined at various angles to achieve forward motion. Sequential contraction occurs, with each myotome contracting after that in front. Cephalochordate muscle fibres possess muscle tails that project from the myotomes to the nerve cord.

c Muscles of two types exist with features as shown in Table 33.1.

Table 33.1

Type	Response	Rise time msec	Resting Potential	Action Potential
FAST	Twitch	70—80	ca50mV	30—70mV
SLOW	Twitch	200—250	ca50mV	1—10mV

Phylum CEPHALOCHORDATA

Co-ordination

a *i* The central nervous system is a dorsal, hollow nerve cord which is specialized anteriorly to form a brain. 'Giant' fibres run in the trunk cord, and there are also giant multipolar cells (Rhode cells). Dorsal roots lie between the myotomes and ventral roots lie opposite myotomes.

Sense organs are limited in nature and extent. Anteriorly there is an olfactory pit opening on the left side. Tactile receptors exist on the oral cirri. Cells within the nerve cord named after Retzius and

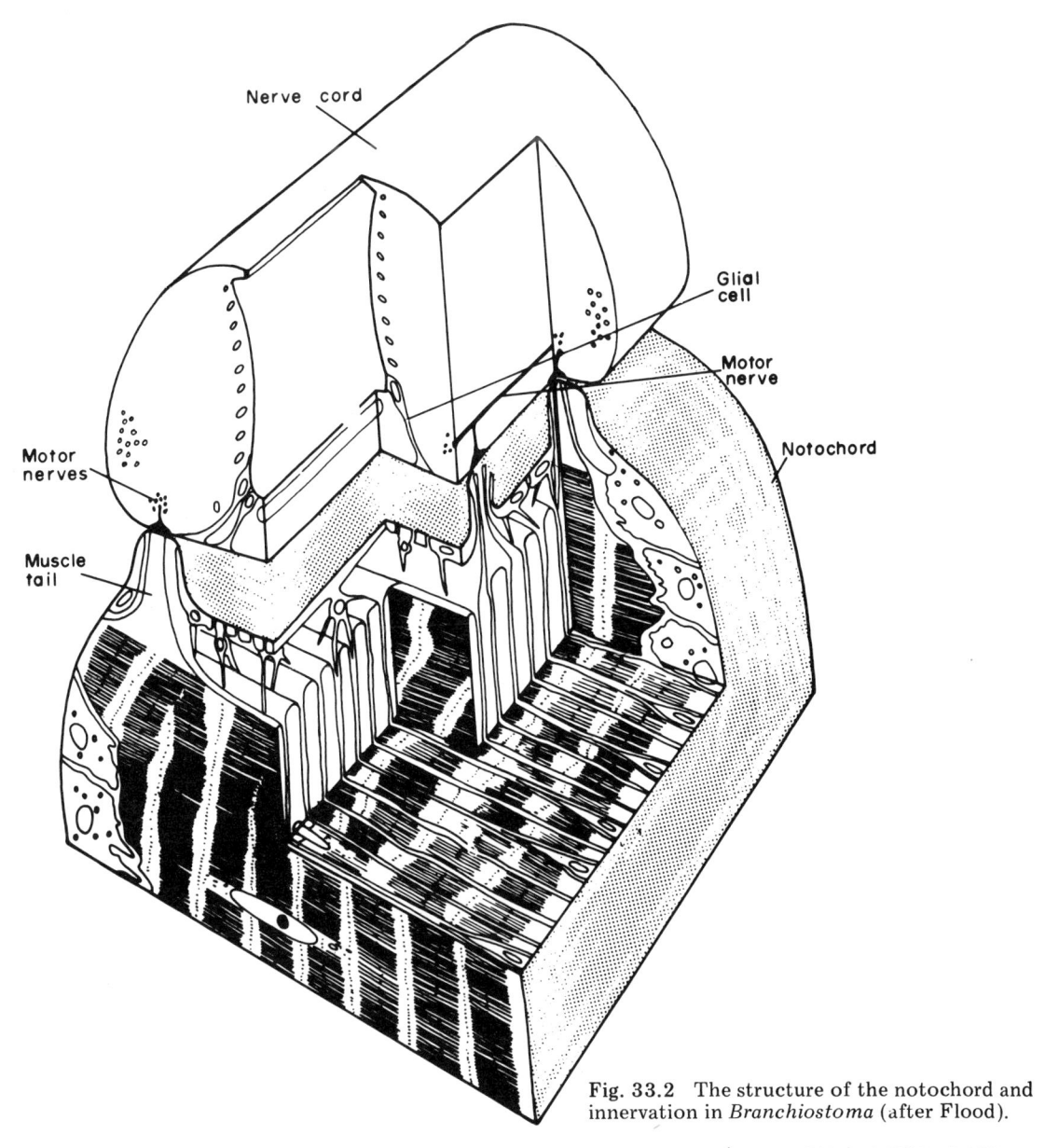

Fig. 33.2 The structure of the notochord and innervation in *Branchiostoma* (after Flood).

Phylum CEPHALOCHORDATA

Joseph are believed to be sensory, the latter photosensory.

ii Stimulation of the ventral region of the nerve cord in the pharyngeal area leads to ipsilateral contraction of the myotomes or powerful contraction of the pterygial region (floor of pharynx). Stimulation in the post-atrial portion of the ventral nerve cord invokes rhythmic contraction involving myotomes of both sides.

b Adult hormones are not well known although the endostyle binds iodine and synthesises mono-, di- and tri-iodothyronine. The metamorphosis of the larva to the adult is triggered by contact with the bottom and is though to be controlled by the club-shaped gland which appears in the pharynx, lasts through the juvenile stage and then regresses.

Respiration

a This is presumably the function of the gill bars forming the wall of the pharynx, which is perforated by many gill slits (pharyngotremy). Each primary gill bar contains a continuation of the coelomic space, and all gill bars (primary and secondary) are served by blood vessels.

b Water enters the mouth, passes into the pharynx, moves through the gill slits into the atrium, and is voided via the atriopore. Gaseous exchange must occur across the ciliated surfaces of the gill bars.

Osmoregulation/excretion

a The organs are solenocytes (flame cells). These are located on each primary gill bar above the pharynx. The solenocytes are closed internally but open to the atrium via a sac, into which many cells void their filtrate.

b The manner in which these cells function is not known.

Circulation/coelom

a The water-filled atrium is large and the coelom is much restricted by its presence anteriorly. Two anterior cavities are found dorsally in the roof of the pharynx, connecting with a median, ventral cavity via channels in primary gill bars. Further to the rear the coelom surrounds the intestine but is reduced on the right side.

A closed vascular system carries blood around the body. A ventral aorta carries blood forward, from whence it escapes along vessels in the gill bars to the paired dorsal aortae. These fuse to a single, posterior, dorsal aorta in which blood flows backward. Branches carry blood through intestinal vessels to a ventral sub-intestinal vessel, and thence foward via the hepatic region to the ventral aorta.

b Propulsion is achieved by pulsations in the ventral aorta, by contractions at the bases of the branchial arteries, and possibly by muscular contractions of the body wall.

c The blood is colourless, has a few red corpuscles and permeates lymph spaces in the fin ray and lateral fin regions.

Reproduction

a There are separate sexes. The gonads are located serially on the pharyngeal wall and open directly into the atrium.

b Fertilization is external. The eggs are not very yolky. Development gives rise to a larva.

References

Barrington E.J.W. 1965. *The Biology of Hemichordata and Protochordata.* University Reviews in Biology. Oliver & Boyd, Edinburgh and London.

Guthrie D.M. & Banks J.R. 1970. Observations on the function and physiological properties of a fast paramyosin muscle — the notochord of Amphioxus (*Branchiostoma lanceolatum*). *J. Exp. Biol.* **52**, 125—138.

A Systematic Index

Acanthocephala 2, 63
Acoela 35, 39
Afrenulata 97
Amphineura 140, 141, 143, 145, 146, 148, 149
Anasca 160
Annelida 2, 4, 5, 74, 90, 91, 93, 96, 100
Anopla 45
Anthozoa 23, 25, 27, 28
Aplacophora 140, 149, 151
Appendicularia 180
Apterogota 106
Arachnida 105, 107, 108, 112, 113, 114, 115, 116, 117, 120, 124, 125, 126, 128, 129
Archaeogastropoda 141
Archiannelida 76, 100
Areneida 119
Arthropoda 2, 4, 102, 105, 106, 125
Articulata 162
Ascidiacea 179, 180, 181, 183
Asconoid 18, 19
Ascophora 160
Aspidogastrea 34, 37
Asteroidea 168, 169, 170, 171, 172, 173, 174, 175, 176, 178
Asterozoa 171
Athecanephria 97

Basommatophora 142
Bdelloidea 55
Bdellonemertini 45
Belemnoida 144
Bivalvia 150
Brachiopoda 2, 162
Branchiopoda 105, 117
Bryozoa 157

Calcarea 19
Cephalocarida 105
Cephalochordata 2, 5, 189
Cephalopoda 15, 138, 142, 144, 145, 146, 148, 149, 150, 151, 152

Cestoda 2, 34, 35, 36, 37, 39, 40, 41, 42
Chaetognatha 2, 166
Chaetonotoidea 50
Chilopoda 105, 110, 111, 112, 115, 116, 117, 119, 124, 126, 127, 129, 130
Ciliata 6, 11, 14
Ciliophora 6, 9
Cirripedia 105, 130
Cladocera 130
Cnidaria 1, 21, 22 25
Cnidospora 10
Coelenterata 4
Coleoidea 143, 144
Comatulida 170
Copepoda 105, 117, 130
Crinoidea 168, 169, 170, 171, 172, 173, 175, 176, 178
Crinozoa 170
Crustacea 5, 105, 108, 109, 112, 113, 114, 115, 117, 118, 120, 123, 124, 125, 126, 128, 130
Ctenolaemata 160
Ctenophora 1, 31
Cubomedusa 28, 29
Cydippid 32

Decapoda 109, 110, 120, 125, 130
Demospongiae 18, 19
Deuterostomia 3, 4, 97, 100, 187
Dibranchia 149
Dicyemida 15, 17
Digenea 34, 37, 41, 42
Diplopoda 105, 109, 112, 116, 117, 124, 126, 127, 128, 130
Diptera 123
Doliolidae 179, 181, 183

Echinodermata 1, 2, 4, 168
Echinoidea 168, 169, 171, 172, 174, 175, 176, 178
Echinozoa 171
Echiurida 87, 88, 91

Ectoprocta 2, 157
Endopterygota 106
Enopla 45
Enteropneusta 186, 187, 188
Entoprocta 2, 71
Eulamellibranchiata 145
Exopterygota 106

Flagellata 6, 13
Foraminifera 11
Frenulata 97

Gastropoda 139, 140, 141, 143, 144, 145, 146, 148, 150, 151
Gastrotricha 2, 49, 59
Gregarina 10
Gymnolaemata 160

Heliozoa 11
Hemichordata 2, 4, 5, 185
Hemimetabola 106
Heteronemertini 45, 46
Heteronereis 82, 85
Heteropoda 146, 147
Hexactinellida 19
Hirudinea 74, 78, 84
Holometabola 106
Holothuroidea 168, 169, 171, 172, 173, 174, 175, 176, 178
Hoplonemertini 45, 46, 47
Hydrozoa 23, 24, 27, 28, 29, 30
Hypermastigina 12

Inarticulata 162
Insecta 5, 103, 106, 112, 113, 114, 115, 116, 117, 118, 119, 120, 122, 123, 124, 125, 126, 127, 129, 130
Isopoda 109, 111

Kinorhyncha 2, 50, 52, 59

Lamellibranchia 138, 142, 144, 145, 146, 148, 149, 151, 164
Larvacea 180, 183
Leeches 80, 84
Leuconoid 19, 20
Limicolae 77
Linguatulida 135

Malacostraca 115
Mastigophora 6, 7
Megascolecidae 77
Merostomata 105, 107, 112, 116, 117, 123, 124, 125, 126, 128, 129
Mesogastropoda 141, 150
Mesozoa 1, 15
Metazoa 1, 2, 13
Mollusca 2, 5, 100, 138, 164
Monogenea 34, 37, 38
Monoplacophora 140, 143, 146, 148, 150
Myriapoda 118, 119, 120, 123
Mystacocarida 105

Nautiloidea 142
Nematoda 2, 66
Nematomorpha 2, 60
Nemertini 2, 44
Neogastropoda 141, 150
Nereid 83
Neritacea 150
Nuda 31

Oligochaeta 74, 77, 80, 82, 83, 84
Onychophora 102
Opalinatea 6, 9

Ophiuroidea 168, 169, 171, 172, 173, 175, 176, 178
Opisthobranchia 142, 146, 149
Orthonectida 15
Orthoptera 120

Palaeonemertini 45, 46
Pauropoda 105, 110, 112, 116, 125, 126, 129, 130
Pentastomida 135
Phoronida 2, 153
Phylactolaemata 159, 161
Phytoflagellata 6, 7, 10, 13
Plasmodiidae 11
Platyhelminthes 2, 34
Pogonophora 97
Polychaeta 75, 80, 82, 83, 84, 85, 87, 100
Polycladida 39
Polyplacophora 145, 150
Porifera 1, 18
Priapuloidea 2, 57
Prosobranchia 139, 140, 144, 148, 149, 150
Protobranchia 142
Protostomia 3, 4, 97
Protozoa 1, 6, 21
Pterobranchiata 186, 187, 188
Pterygota 106
Pteropoda 146, 157
Pulmonata 142, 146, 150, 151
Pycnogonida 105, 108, 112, 113, 116, 117, 125, 126, 128, 129
Pyrosomidae 181

Radiolaria 1, 11
Rhizopoda 1, 6
Rotifera 2, 54, 59

Salpidae 179, 181, 183
Sarcodinea 6
Scaphopoda 142, 144, 145, 146, 148, 150, 151
Scorpionida 115
Scyphistoma 22
Scyphomedusae 28, 29
Scyphozoa 24, 27, 29
Sepioida 144
Septibranchia 142
Siphonophora 30
Sipunculida 91, 93, 96
Solenogastres 140
Solifugae 115
Spongia 4
Spongillidae 18
Sporozoa 6, 10, 11, 12, 13
Stylommatophora 142
Suctoria 11
Syconoida 19, 20
Syllida 83
Symphyla 111, 112, 115, 116, 124, 126, 127, 129, 130

Tardigrada 132
Tentaculata 31
Terricolae 77
Testacea 11
Tetrabranchia 149
Thaliacea 179, 180, 181, 183
Thecanephria 97
Trilobita 107

Urochordata 2, 5, 179

Xiphosura 107, 117

Zooflagellata 6, 7

Index of Genera

Acanthobdella 78, 79
Acanthocephalus 63, 64
Acanthogyrus 63
Acanthometra 6, 8
Acineta 9
Actinia 25
Actinophilus 10
Actinosphaerium 6, 8
Aeolidia 141
Aeschna 111
Agelena 130
Alaria 35
Alcyonidium 160
Alcyonium 25
Allocreadium 37
Alloteuthis 142
Alma 82, 83
Amoeba 6, 8, 11, 12
Amphiporus 45
Amphitrite 75, 76
Amphiura 175
Anaspides 111
Ancylostoma 66
Androctonus 107
Antedon 170, 175
Aphrodite 82
Apis 111
Aplysia 147
Araneus 107
Arenicola 75, 82, 83
Argonauta 142, 152
Arion 142
Artemia 108
Ascaris 66, 69
Aspidogaster 37
Astacus 110, 115
Asterias 171, 175
Asterina 178
Astropecten 171, 175
Asymmetron 189
Aurelia 22, 24

Balanoglossus 186
Barentsia 72
Belosepia 144

Beroë 31
Birsteinia 99
Bonellia 87, 88, 89, 91
Botryllus 180
Bougainvillea 23
Branchiodrilus 82
Branchiostoma 189, 190, 191
Buccinum 141
Bugula 158

Calanus 108
Calliobothrium 40
Cambarus 109
Campylodeses 52
Cancer 108
Carinoma 45
Cassiopeia 24
Cephalobaena 135, 136
Cephalodasys 49
Cephalodiscus 186, 187
Ceratium 6, 7
Cerebratulus 45, 47
Cestum 31
Chaetonotus 49, 50
Chaetopterus 82
Chaos 13
Chlamydomonas 13
Cidaris 171
Ciona 180
Clavelina 180
Clione 147
Codosiga 7
Coeloplana 31, 33
Conoteuthis 144
Conus 144
Crania 162
Crepidula 144, 151
Crisia 160
Cristatella 159
Crossaster 171
Cryptoplax 140
Cryptomonas 6
Ctenoplana 31, 33
Cucumaria 171, 178
Cuspidaria 142
Cyanea 24

Cyclops 109
Cyclopora 160
Cypris 108

Daphnia 108
Dendrocoelum 35, 37
Dendrocometes 6
Dendrostomum 93, 94, 95
Dentalium 142
Dermacentor 108
Dero 82
Diclidophora 37
Diclidophoropsis 38
Dicyemida 15
Difflugia 6, 8
Diphyllobothrium 36, 37
Discona 162
Doliolum 180
Doris 142
Dytiscus 111, 115

Echinocardium 171, 173, 175
Echinococcus 37
Echiniscus 132
Echinocyamus 171
Echinoderella 52
Echinoderes 52, 53
Echinus 171, 172, 175
Echiurus 87, 88, 89
Eimeria 6, 10
Eisenia 77
Electra 157
Eledone 142
Ensis 142
Entamoeba 6
Eolis 142
Eoperipatus 104
Epiphanes 54
Epithetosoma 90
Erpobdella 84
Euchlora 31
Eukrohnia 166
Euglena 6, 7
Eunicella 25
Euplectella 19

Eutonina 23
Eutyphoeus 83

Fasciola 37
Forficula 111
Fritillaria 180
Fungia 25

Geonemertes 46
Geophilus 110
Geryonia 23
Globigerina 8
Glomeris 109
Glossina 111
Glossiphonia 78
Glossoscolex 83
Glottidia 165
Glycera 83
Gordius 60
Gorgonocephalus 171, 173
Gyrodactylus 37

Haemopis 78, 84
Halichondria 19
Haliclystus 24
Halicryptus 57, 59
Haliotis 141, 149, 151
Haplosporidium 6
Helicosporidium 6
Helix 141, 142, 147
Henricia 172
Heterakis 66
Hirudo 78, 80, 81, 82, 84
Holothuria 171
Homarus 108, 126
Hyalonema 19
Hydatina 54
Hydra 23, 27, 29
Hydractinia 23
Hymenolepis 36

Ichthyosporidium 10
Ikeda 87, 89
Ixodes 107

Lamellaria 138
Lamellibrachia 97
Lamellisabella 98, 99, 101
Lepidochiton 140
Lepidoderma 49
Lepisma 111
Leptasterias 178

Leptosynapta 178
Leucosolenia 19
Ligidium 109
Limnaea 142
Limulus 107, 115, 125
Linckia 178
Lineus 45
Linguatula 135, 137
Lingula 162, 164
Littorina 141
Locusta 111
Loligo 143, 144, 147
Lophomonas 7
Loxosoma 71, 72
Lucernaria 24
Lumbricus 77, 80, 82, 83, 84
Luidia 171

Macrobiotus 132
Malacobdella 45, 46, 47
Megascolex 74
Melanoplus 113
Membranipora 157, 160
Menthus 114
Mertensia 31
Mesostoma 35
Metridium 25
Mnemiopsis 31
Molgula 180
Monas 6, 13
Moniliformis 63
Monocystis 6, 10
Mopalia 140
Mysis 108
Mytilus 142, 146
Myxicola 81
Myxobolus 10

Nautilus 140, 142
Nectonema 60, 61
Neopilina 140, 141, 149, 150, 151
Nephthys 82
Nereis 75, 82
Nerilla 76
Nerillidium 76
Neuronemertes 47
Nosema 10
Nucella 141
Nucula 142
Nymphon 108, 109

Obelia 23
Octopus 143, 145, 147, 150, 151
Oikopleura 180

Oligobrachia 97, 99
Ooperipatus 104
Opalina 6, 9
Ophiocomina 171
Ophiothrix 171
Opisthorchis 37
Orthonectida 15

Pandorina 7
Papilio 111
Paracaudina 175
Parameoium 6, 9, 12, 14
Parorchis 35
Patella 141, 144
Pauropus 110
Pecten 142
Pedicellina 71, 72
Pediculus 111
Pelagothuria 175
Peripatopsis 102, 104
Peripatus 102, 103, 104
Periplaneta 111, 113
Phascolion 93
Phascolosoma 93, 94
Pheretima 77, 80
Philodina 55
Pholas 142
Phoronis 153, 154, 155, 156
Phoronopsis 153
Phyllodoce 75
Plakina 19
Plasmodium 6, 11
Platynereis 83
Pleurobrachia 31
Pleurobranchea 147
Plumatella 159
Polybrachia 99, 101
Polycelis 35
Polydesmus 109
Polygordius 76, 85
Polystoma 37
Pomatoceros 75
Pontobdella 78
Porania 172
Porites 25
Porocephalus 135
Priapulus 57, 58, 59
Proneomenia 140
Protodrilus 76
Pseudicyema 15
Pterotrachea 147
Ptychodera 186
Pulex 111
Pycnogonum 108
Pyrosoma 180

Rhabdopleura 186, 187
Rhizocrinus 170
Rhopalura 15, 16
Rhynchodemus 46

Sabella 75, 82
Sabellaria 74
Saccocirrus 76
Saccoglossus 186
Sagitta 166, 167
Salpa 180, 183
Schistosoma 37
Sclerolinum 101
Scolopendra 111, 118
Scutigera 110
Scutigerella 111, 130
Scyphistoma 22
Sepia 144, 146
Sepiola 146
Serpula 84
Siboglinum 97, 98, 101
Sidnyum 180
Sipunculus 93
Spadella 166, 167
Spirobrachia 99

Spongilla 19
Squilla 108
Stephanoceros 54, 55
Stentor 6, 12
Stomphia 27
Strongylocentrotus 175
Stylaria 77
Stylonychia 9
Sycon 19
Symphylella 111
Synapta 171

Tachypleus 107
Taenia 37, 39, 40
Tealia 25
Temnocephala 35
Terebella 79, 82
Terebratulina 162
Thalassema 88, 89
Tjalfiella 33
Tomopteris 76
Travisia 83
Trichomonas 7
Trichonympha 6
Trypanosoma 6, 7, 12

Tubifex 77
Tubipora 25
Tubulanus 45

Uca 126
Urechis 87, 88, 90
Urnatella 71
Urodacus 108

Vallicula 33
Velamen 31, 32
Velella 23
Viviparus 141
Volvox 6
Vorticella 6, 9, 12

Waddycephalus 136

Yoldia 142

Zelleriella 6

Kirtley Library
Columbia College
8th and Rogers
Columbia, MO. 65201